Ralph Abercromby

Three Essays on Australian Weather

Ralph Abercromby

Three Essays on Australian Weather

ISBN/EAN: 9783337337001

Printed in Europe, USA, Canada, Australia, Japan

Cover: Foto ©berggeist007 / pixelio.de

More available books at **www.hansebooks.com**

THREE ESSAYS

ON

AUSTRALIAN WEATHER

HON. RALPH ABERCROMBY,

Fellow of the Royal Meteorological Society, London; Member of the Scottish Meteorological Society; and Author of "Principles of Forecasting by Means of Weather Charts."

Sydney:
FREDERICK W. WHITE, PRINTER, 39 MARKET STREET WEST.
1890.

INTRODUCTION.

THE following essays on "AUSTRALIAN WEATHER" have been brought together and published in book form by the Hon. RALPH ABERCROMBY, for convenient reference by students of Meteorology, because he thinks that they are important contributions to our knowledge of the Meteorology of a new field. In passing it may be mentioned that the preparation of them has been made possible by the very complete Meteorological records of the Sydney Observatory, especially the continuous automatic records of wind, rain, barometer, and weather charts introduced by the present Director. These afford the means of reference to wind and weather at any moment since 1863, and the barograph since 1870, without which it would have been impossible to study in the necessary detail the so-called "Southerly Burster," which is perhaps the most remarkable of the "squall" winds which are found in various parts of the earth.

The first essay, "Moving Anticyclones," is here reproduced with the consent of the author. It establishes the progressive easterly motion of weather over Australia. The great importance and controlling influence of anticyclones upon the usual weather, traces their progress over Australia, and the changes produced in them by the contour of the country, gives records of the number of such weather systems which pass over Australia every month and year, and traces their motion in latitude with the seasons, and shews that their average daily progress over the mainland is four hundred miles per day, and produces evidence which seems to justify the conclusion that they travel rather faster over the ocean between South Africa and Australia, that is, at the rate of four hundred and fifty-eight miles per day. Twenty diagrams of weather and anticyclone tracks are used in illustration.

The Prize Essay on "Southerly Bursters" was the outcome of a valuable prize offered by The Hon. RALPH ABERCROMBY. It follows the "burster" through all its phases, its connection with moving anticyclones, with the state of the barometer, with the heat of the interior, and with various other phases of weather. Traces them historically, and gives reasons for thinking that they are not so violent now as they used to be. It is illustrated by four cloud pictures and eleven diagrams.

"Types of Australian Weather." These types forming the third essay, were selected by Mr. ABERCROMBY, and he marked out the way they should be treated, and engaged Mr. HUNT to do the actual work necessary for their publication. These studies, as well as those in the essay on "Moving Anticyclones" have been rendered possible by the extreme care with which the Sydney daily weather chart has been prepared, and passing phases of weather submitted to special study since its commencement. Forty of the weather maps on a reduced scale are used in illustrating the types.

MOVING ANTICYCLONES IN THE SOUTHERN HEMISPHERE.

By H. C. Russell, B.A., F.R.S., F. R. Met. Soc.

Sydney Observatory, New South Wales.

[With Twenty Diagrams.]

[*From the Quarterly Journal of the Royal Meteorological Society, Vol. XIX. No. 85. January, 1893.*]

Since Western Australia established in 1887 a number of new Meteorological Stations, it has been possible to make our Weather Charts more complete and to trace the progress of meteorological conditions more minutely : and the opportunity has not been lost. Some of the results are of purely local value as aids in forecasting. Others seem to have more general significance, and will, I think, be of interest to the Fellows of this Society and to meteorologists generally, because they have an important bearing upon accepted theories in explanation of " weather " in latitude 20° to 50° South. So much so that I think it will be necessary to modify those theories. There can be no doubt that the great extent of ocean, as compared with the land, in the latitudes named, affords to atmospheric circulation a field in which it may approach what it would be, if the earth were completely covered by water, and the atmosphere therefore free from the disturbing influences of unequal heating, surface land friction, and mountains.

The leading fact that our[1] investigations have brought to light is that Australian weather south of 20° South latitude is the product of a series of rapidly moving anticyclones, which follow one another with remarkable regularity and are the great controlling force in determining local weather.

---

[1] In the investigations which lead up to the results detailed in this paper I have been very ably assisted by Mr. H. A. Hunt, who prepares the daily Weather Chart, and who has carried out many investigations to the successful discovery of weather laws here.

A

The fixed anticyclone over the Indian Ocean in these latitudes, which is found in books of reference, must give place to a moving series. It is not difficult, now that the true weather conditions are known, to trace the development of the idea of a fixed anticyclone over this part of the ocean. The moving anticyclones are about five times as large in area as the low pressure ʌ between them, and they are also moving to the east at the rate of about 400 miles per day. Now vessels crossing these at random would necessarily find five times as much high pressure as low pressure: and if travelling with the anticyclone would keep in it for many days, perhaps all the way from the Cape to Sydney, as the *Havanah* did. When all these barometer readings came to be plotted on the chart the result would be a fixed anticyclone, which we now know has no existence in fact. It is not so easy to trace the history of the "overland" fixed anticyclone of more modern writers, but it is equally imaginary. The high pressures regularly move on over land as well as over ocean.

It will perhaps be more convenient if I state here, in as few words as possible, some of the results obtained, and bring forward subsequently in more detail the data upon which they are based. These results may be briefly stated as follows:—

1. Instead of fixed anticyclones we have a series of moving ones.

2. The average number of anticyclones passing over Australia in a year is forty-two, and so far as the observations go, the number varies but little. See Table I. (p. 5).

3. Anticyclones are more numerous in summer than in winter. See Table I., and diagrams 9 to 20.

4. The latitude of anticyclone tracks varies with the season, being in latitude 37° to 38° in summer, and 29° to 32° in winter. See Table II. (p. 7).

5. Upon the average, an anticyclone travels across Australia in seven days in summer, and in nine days in winter. Since forty-two pass over each year, the average time of passage over any place is 8·7 days.

6. The average daily rate of translation derived from all the available records is, over Australia, four hundred miles; and over the sea and land from Natal to Sydney, four hundred and fifty-eight miles. (See page 8.) The rate of translation of moving anticyclones varies from two hundred to nine hundred and fifty miles over land.

7. The shape of the anticyclone over the comparatively flat lands of Australia is an ellipse, with axes in ratio of two to one, the longer axis being east and west. The shape, as well as the direction of the major axis, are, as a rule, modified when the anticyclone reaches the coast range, the result being a shortening of the major axis, and a bending of the major axis to or towards a position at right angles to what it had on the low lands, *i.e.*, making it north and south. See diagrams 1, 2, 3, 4, 5, 6, 7, 8 and 9.

8. The winds on the north side of the anticyclone are not so strong as those on the south side; but at the ends the winds have greater velocity, and the two winds, northerly and southerly, pass each other as if struggling to get through between two obstacles, *i.e.* the preceding and following anticyclones. See diagrams 1 to 8.

9. The intensity of weather is in proportion to the difference in pressure between the anticyclone and the A depression, but the relation of the pressure varies frequently *before* the wind responds, and it seems as if the pressure was controlled from above by the more or less rapid descent of air, which feeds the anticyclone. The centre of the descending current, assuming that to be the point of greatest pressure, not unfrequently moves about independently of the general motion, so that the centre at times seems to retreat, without corresponding motion in the extreme parts of the anticyclone.

10. When the A depression is deep it is usual for the South-east Trade wind, blowing in the north of Australia, to be deflected into a Northerly wind in the rear of an anticyclone, which gets

heated as it blows over Central Australia, and becomes the true hot wind of the Southern Colonies. This also explains the sudden shift from hot northerly to the cold southerly wind, blowing on the following side of the ʌ depression.

11. Cyclonic storms are very unusual, and do not appear more than once in two or three months. These come from north-east to east, or from the north-west coast across the Australian continent to the sea at the Great Bight, thence they travel eastward. They do not seem in any sense to be part of our weather system, but to be offsets from tropical storms.

The depression between anticyclones is essentially a ʌ depression, both in the shape of the isobars and in the sudden change of wind which follows the passage of the lowest pressure.

It is, of course, impossible here to pass in review all the 1,400 Weather Charts which have contributed to these deductions; and I have, therefore, selected for reproduction here a set of eight Weather Charts in which the passage of an average cyclone is clearly depicted (Diagrams 1 to 8). The isobars have been reproduced on a convenient and small scale, and they show much better than any description what the ordinary fine weather sequence of events is. Attention may, however, be called to one or two points. No. 1 diagram shows the incoming anticyclone well established on the coast of Western Australia, with a passing ʌ depression in front of it extending over the southern parts of South Australia, Victoria, and New South Wales. No. 2 shows the twenty-four hours' forward motion and the closing up of the isobars as they reach the east coast mountain range. Nos. 3 and 4 show the further progress of the high pressure and the decided effect of New Zealand mountains in intensifying the ʌ depression. No. 5.—As on Sunday we get no telegrams, *probable* isobars have been drawn. No. 6 shows the anticyclone over New South Wales, and the preceding end of its major axis tilted northwards by the mountains, while in Western Australia we see the first isobar of the incoming high pressure, and over Perth the ʌ depression which divides the two. No. 7.—The axis of the anticyclone is

MOVING ANTICYCLONES.

now nearly north and south, and it has passed over the mountains. No. 8.—The anticyclone is over the sea between Australia and New Zealand, and with its two axes nearly equal. Such is the passage of an ordinary anticyclone. This one was moving at the rate of four hundred and fifty miles per day, which is somewhat above the average speed; and by referring to the diagrams it will be seen that its track is nearly straight, and from west to east. The majority of such anticyclone tracks are bent southwards in the middle, as if when the anticyclone reaches the west coast range, which trends to south-south-east, it is deflected so as to move to the east-south-east instead of east; and that when it meets the east coast range which trends to north-north-east the track is again deflected by mountains and made to go east-north-east or north-east. See illustrations of this in diagrams 11, 15, 17, 18, 19, and 20.

Another feature brought out in the diagrams is the occasional stoppage of the anticyclone. For instance, in diagram 12, track C, the centre was about the same place from April 25th to May 1st. In No. 15 the track is remarkable, the anticyclone lasting twenty-five days, although it moved every day. Of its kind, this is the most remarkable one we have on record.

Out of a total of forty-two anticyclones which passed over Australia in 1891, six, or fifteen per cent., hesitated or actually stopped in their forward motion. Table I. shows the number of cyclones in each month since February 1888.

TABLE I.—NUMBER OF ANTICYCLONES IN EACH MONTH.

Years.	January.	February.	March.	April.	May.	June.	July.	August.	September.	October.	November.	December.
1888......	...	...	2	3	3	2	3	5	3	5	4	4
1889......	3	4	2	3	4	4	3	3	4	4	4	5
1890......	5	3	4	3	2	2	2	3	2	5	5	6
1891......	4	4	3	3	3	4	2	4	4	4	4	2
1892......	3	4	3	4	3	4	3	...	...	...	...	...
Totals.....	15	15	14	16	15	16	13	15	13	18	17	17
Mean......	3·8	3·8	2·8	3·2	3·0	3·2	2·6	3·8	3·3	4·5	4·3	4·3

It is difficult to understand how an anticyclone can, to all intents and purposes, stand still for several days when the whole surrounding atmosphere is moving forward; and we have not yet made out the explanation. But facts, and some very significant ones, are accumulating which indicate the probable explanation. For instance, when such a stoppage occurs, the isobars in front widen out, showing that the preceding system is moving forward; and the closing up of the isobars in the rear shows that the following one is coming forward. We know also, as already pointed out, that the source of pressure, the descending current in the anticyclone, may, and does, vary in locality from day to day with regard to the outlying isobars, coming down at one time on the preceding side of the centre and at the next on the following side. This gives an oscillating position to the apparent centre. But it seems impossible that a mass of air, even that in one anticyclone, without reference to its surroundings, measuring as it does 2,000 miles by 1,000 miles, can be actually stopped in its forward motion. This point is however, one of those still under investigation, and we hope in another paper to make the explanation complete.

In America it is well known that cyclones move ten to fifteen per cent. faster than anticyclones, but there the low pressure acquires a velocity of its own in addition to that which it has in common with the general mass. The conditions here, as already pointed out, are different, the low pressure not being independent, but tied to the anticyclone, as effect is to cause.

Another point of considerable importance is the normal latitude of the anticyclones in each month of the year. At times it is a very uncertain matter, for there is as much difference in the latitudes of tracks in the same month in some cases as there is in the average tracks for each month of the year. Still, taking the four and a half years of available records, there is a very obvious monthly change of latitude, which is best indicated in tabular form.

TABLE II.—MONTHLY CHANGE OF LATITUDE OF THE ANTICYCLONES.

Months.	1888.	1889.	1890.	1891.	1892.	Mean Latitude.
	°	°	°	°	°	°
January	...	37	38	36	36	36¾
February	...	38	41	36	38	38¼
March	34	38	39	37	37	37
April	37	32	35	36	35	35
May	33	32	31	34	33	32⅗
June	27	27	27	37*	29	27½
July	27	30	30	29	33	29⅘
August	28	31	30	30	...	29¾
September	32	32	30	30	...	30½
October	33	32	29	33	...	31¾
November	38	37	32	36	...	35¾
December	39	37	35	35	...	36½

* Not included in mean.

It must be borne in mind that the tracks are seldom straight, and that at times they are very erractic. The above positions for each month have been obtained from careful eye estimates of the mean positions, and this must be borne in mind in using the table. The tracks for 1891, diagrams 9 to 20, will illustrate what is meant, and, will justify the course adopted in preference to that of taking the measured latitude of a series of points in each curve. The individual curves are too irregular for such a method.

It will be observed that the maximum of latitude is in February, a month after the hottest month; and the minimum in June, which is not our coldest month. Perhaps however, June should be rejected from this attempt to determine the monthly latitude, because in that month, for some reason not obvious, the tracks are much more erratic than in any other month of the year, and it is almost impossible to take a mean latitude for it; and if it be rejected the minimum would be between July and August, July being our coldest month.

It is our experience that when an anticyclone track is far from the mean, the weather is also far from the mean. For instance, in June 1891 two tracks were down in latitude 37° instead of 27°, and we had, as a consequence, fourteen and a half inches of rain, an excessive quantity, the mean being five and a quarter inches.

I have already stated, that taking a large number of anticyclones on their passage over Australia the daily translation eastward is four hundred miles ; and it is a matter of considerable interest to ascertain if they maintain the same velocity of translation over the ocean. I have been so far unable to trace any connection between the variations in the barometers at Buenos Ayres and Sydney, and infer from this that the great mountain chain of the Andes so breaks up the anticyclones that the curves are not alike. The only other place in like latitude for which I have daily barometer readings, and these for only one year, 1890, is Natal. Fortunately there is an obvious similarity in the curves. I have compared these with Sydney by having the two plotted on the same scale and taking the difference in the times at which marked points of high and low barometer readings pass the two places. There are—

In January	5	cases with a mean of	13 days
February	5	,,	16
March	5	,,	15
April	3	,,	18
May	4	,,	12
June	3	,,	14
July	4	,,	16
August	5	,,	17
September	5	,,	14
October	4	,,	14
November	3	,,	17
December	5	,,	15
Total	51 cases	Average No. of days	15·08

The greatest number of days for the translation of the waves from Natal to Sydney is eighteen days for April, which makes the daily velocity of translation three hundred and eithty-two miles ; the least number of days in any month is twelve, in May, which makes the velocity of translation five hundred and seventy-three miles per day. The average rate for all the months is fifteen days, which is equal to a velocity of translation of four hundred and fifty-eight miles per day.

Taking individual anticyclones in Australia the slowest moved at one hundred and twenty miles per day, and the quickest nine hundred and fifty miles per day, the average being four hundred miles per day. The result of this comparison surprised me by the similarity of the two results. I do not, however, think that very much value should be attached to a comparison of barometer curves for one year only, still it agrees remarkably with that obtained by four and three-quarter years' results for Australasia, and the little difference is in the direction it ought to be, owing to the obvious friction of the land surface, mountains, etc., as compared with the smooth sea surface.

The two methods of determining the velocity of anticyclones, that is, over Australia alone, where it is four hundred miles per day, and over the space from Natal to Sydney, where it is four hundred and fifty-eight miles per day, seem to leave no doubt as to their persistence. For if they can thus be followed one-third of the circumference of the earth, *i.e.* from Natal to Sydney, it may safely be assumed that they travel the other two-thirds of the way, and that they keep up their general characteristics. What influence that great obstacle in their path, the Andes of South America, may have on them I am not at present in a position to say, but I have no doubt, from what we see so clearly in the influence of our own comparatively small range of mountains along the east coast of New South Wales, that it is a very material one.

If from Buenos Ayres we could get by cablegram the state of the weather from day to day, we should be in a position to forecast the coming weather for about a month in advance; and it may yet be that when our investigations, which are now in progress, are completed we shall be able to forecast far longer periods. If, for instance, we could ascertain the velocity of the translation of the anticyclone round the other two-thirds of the globe, as we have done for the one-third from Natal to Sydney, or rather more than one-third because it extends to New Zealand, then we could ultimately forecast the return, in say, seven weeks, of weather passing over Sydney. Certainly the discovery of the daily translation of anticyclones in our latitude, over such a large section of the circumference of the globe, holds out a reasonable hope that they may be traced all round, and the proportion of water surface points clearly to the fact that the conditions are more favourable here than in any other part of the earth for normal atmospheric circulation. I do not by this intend to convey the idea that I think an anticyclone keeps its shape, size, form, and pecularities for weeks together, because I see them changing every day. But nevertheless there are obvious pecularities which affect some anticyclones—general characteristics I mean, such as dryness or moisture,—which, it may be, are attached to them more persistently than the mere form of the isobars. And if so, it will afford good data for long period forecasting.

If I have succeeded in showing the normal conditions of our weather, to be that of an endless series of anticyclones passing over to the eastward at the average rate of four hundred miles per day and keeping within a very moderate range of latitude, it will be obvious that these conditions hold out the prospect of our being able to predict the weather for some weeks in advance; because an anticyclone with such a rate of motion in latitude 38° south would pass round the earth in forty-nine days as stated above. It is true that in isolated anticyclones we find the rate of motion vary considerably over a part of its track; but I think it is a fair assumption to make, that the average velocity of the anticyclones, in a given latitude, is the rate of progress of the atmosphere at that latitude, and that the apparent variations in rate are simply local accelerations or retardations, which would naturally affect masses of air in motion under several influences, that is, the translation, the constant variations in pressure, temperature, etc. Some of the conclusions arrived at from the examination of four and a half years weather charts, may have to be modified slightly when the investigation is continued over other years; more especially, when the number of observing stations make the data more complete, but the general agreement of the results in the years examined, seems to support the opinion that there will be no material alterations in the results here given.

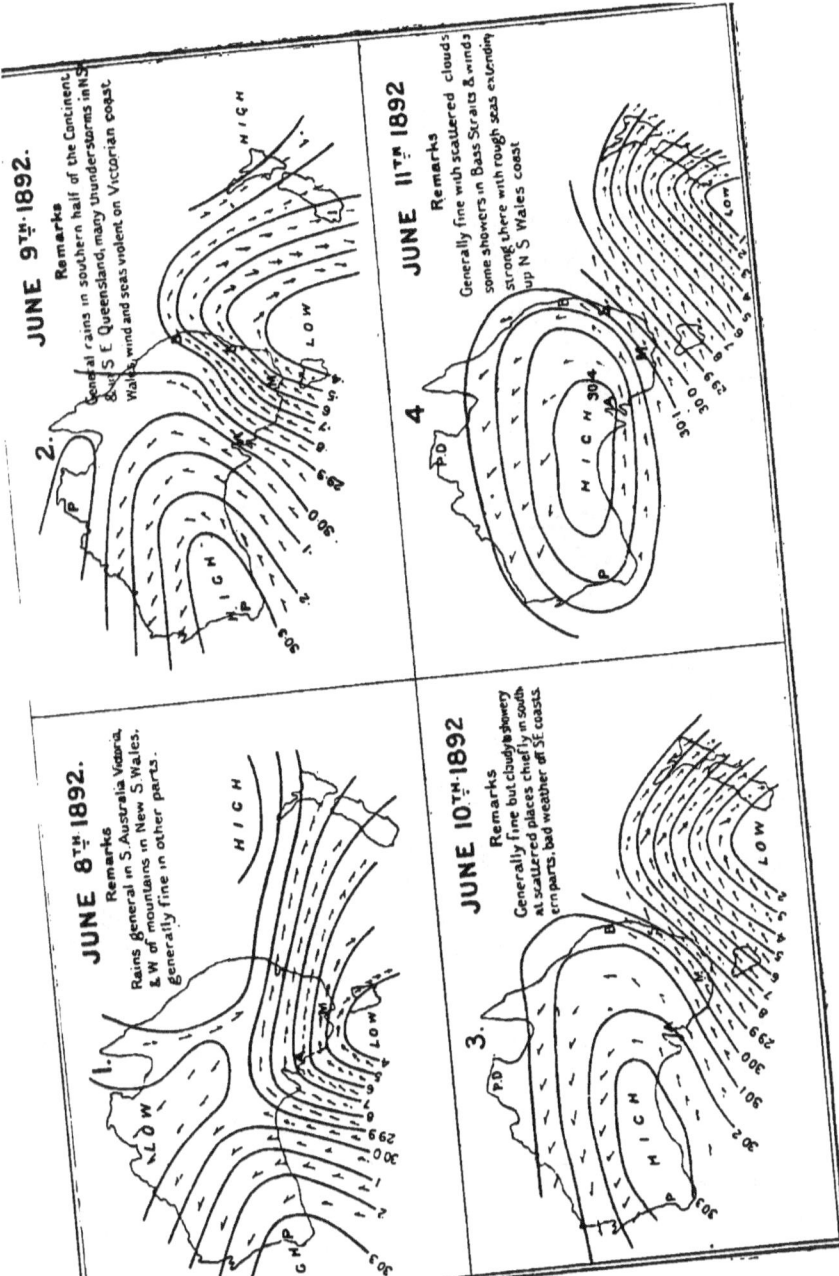

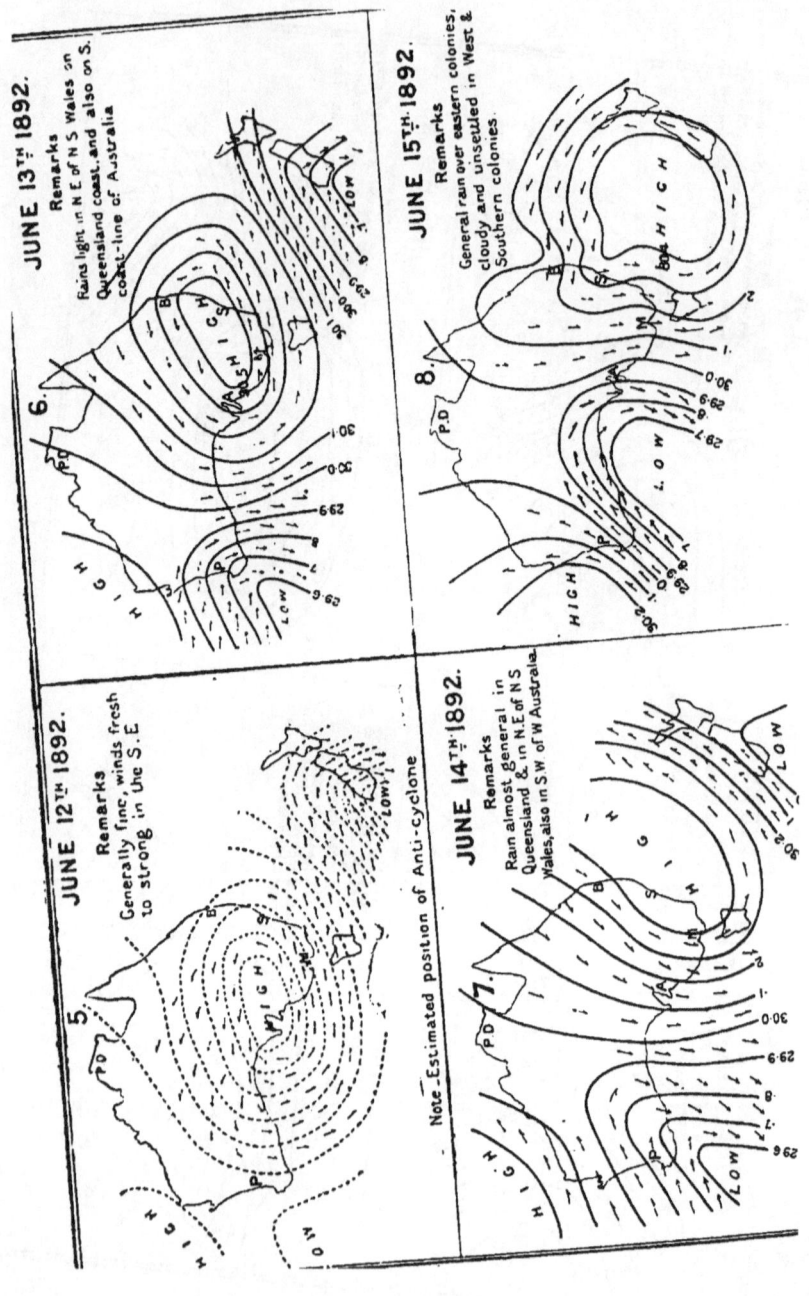

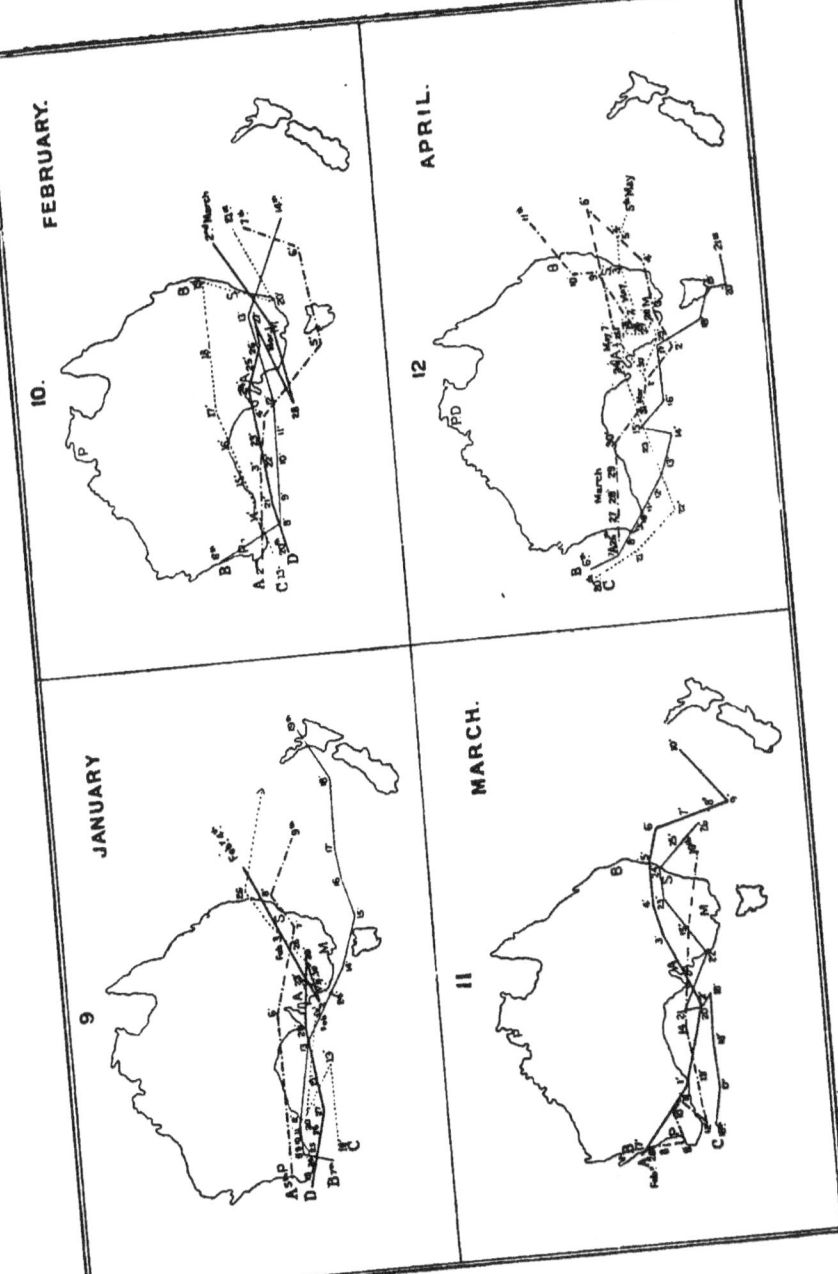

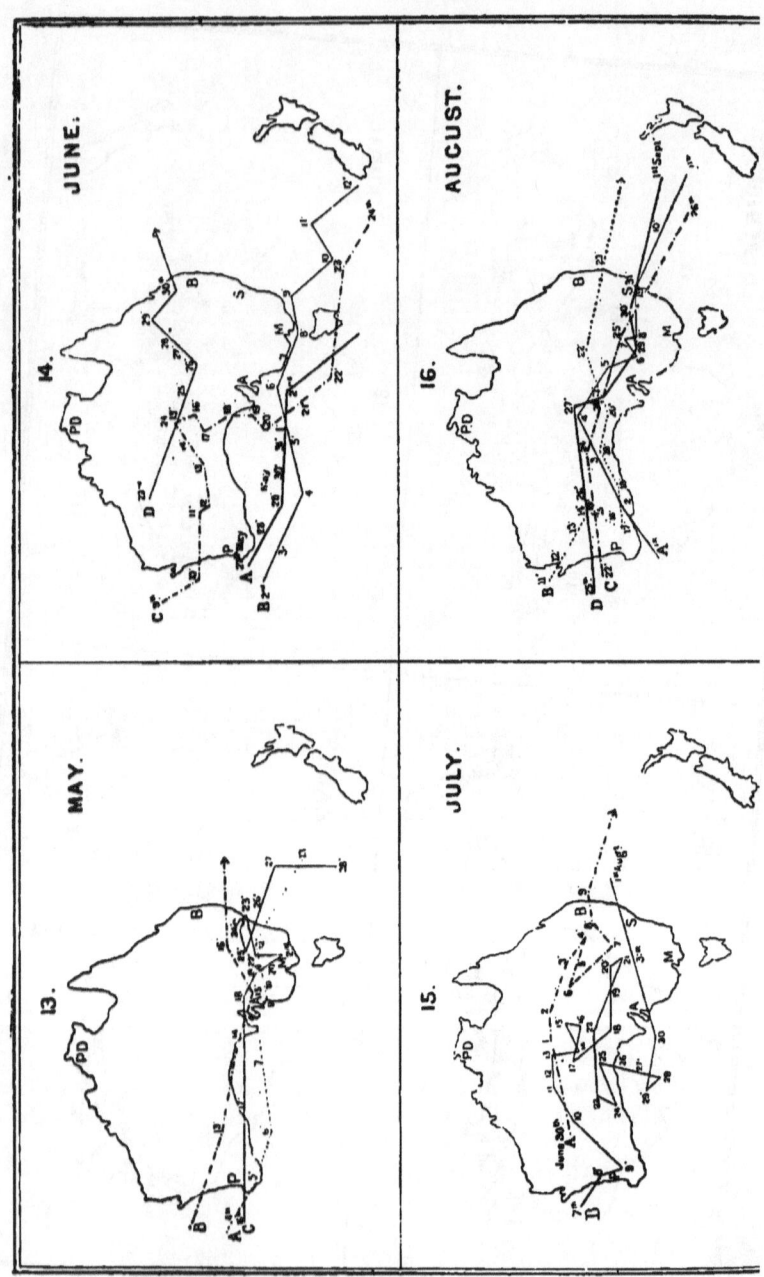

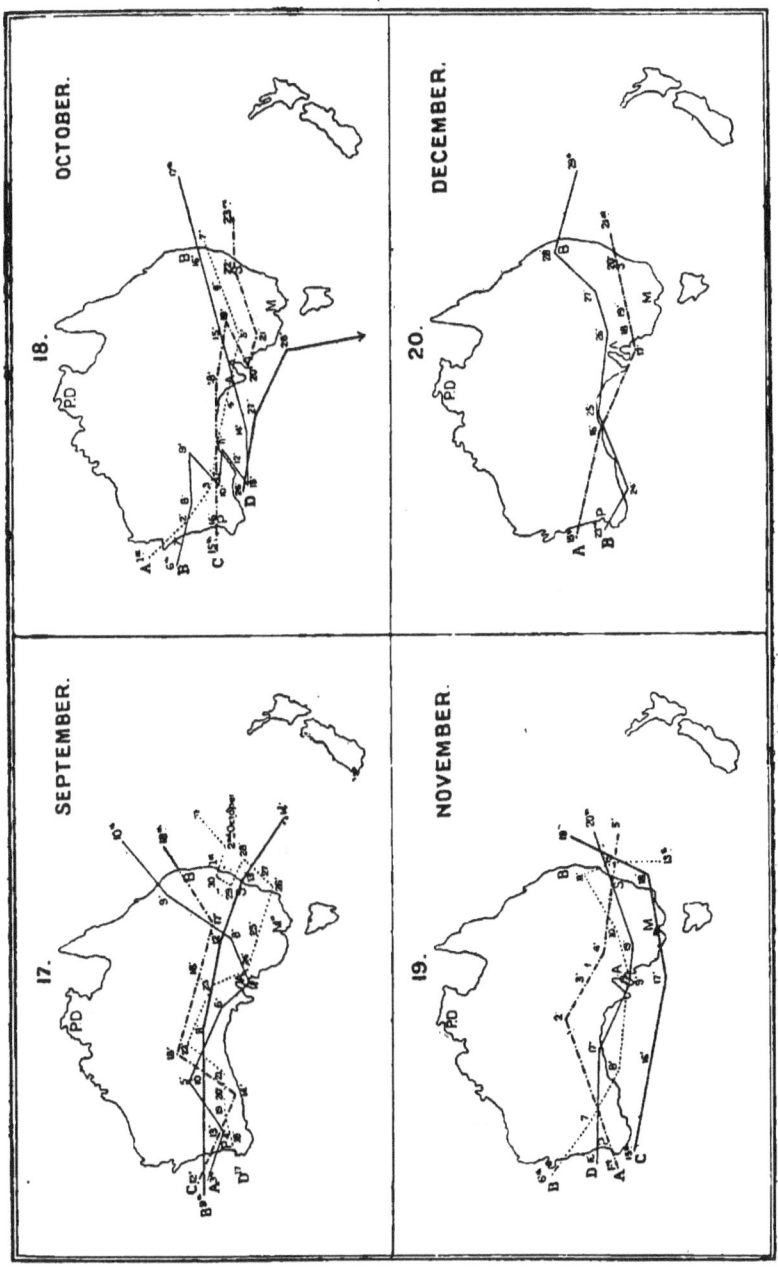

AN ESSAY ON SOUTHERLY BURSTERS.

By HENRY A. HUNT.

Second Meteorological Assistant, Sydney Observatory.

[With Four Photographs and Five Diagrams.]

[Awarded the Prize of £25 offered by the Hon. Ralph Abercromby, for the best Essay on Southerly Bursters, 2 May, 1894.]

ORIGIN ON THE PRIZE.

IN December 1892, the Honorable Ralph Abercromby gave to the Royal Society a sum of £100 to promote the study of Australasian Meteorology by offering prizes for essays upon particular phases of weather, and in rewards for special investigations, suggesting that the subject of the first essay should be "The Southerly Burster." The foundation of this Prize Fund was announced at the general meeting of the Royal Society at the December 1892 meeting. Subsequently the Council appointed a Committee consisting of the Hon. R. Abercromby, Professor Liversidge (Chemistry), Professor David (Geology and Physical Geography), and H. C. Russell, Government Astronomer.

The Committee met to determine the conditions for the competition and advertised these freely in Melbourne and Sydney, and by correspondence with various kindred Societies in Europe and America, and a note appeared in *Nature* about it; one of the conditions was that the competition was open to all. It was advertised in March 1893, and the last day for receiving essays was 31st March, 1894.

The contents of this essay may be briefly summarized as follows: It begins with a short note on bursters past and present, and weather indicating their approach. Deals with the burster in other colonies, shows that it is intensified in New South Wales by geographical features. Is sometimes caused by monsoonal depressions. Traces the changes in isobars with various kinds of bursters. Shows that duration and strength of bursters have wide ranges. Gives a sketch of the Pampero. Traces different kinds of bursters, their rate of progress along the sea coast and relation to general weather conditions. Gives detailed description of two bursters with diagrams of weather before and after, and photographs of clouds, also full notes of cloud changes. Gives diagram and short note about the most violent burster ever known on the coast. Gives tabular particulars of all the bursters that have taken place between September 1863 and March 1894. Showing the number in each hour of the day in each month and in each year, and the total number nine hundred and ninety-one which have been recorded, the prevalence in each month in each year with the greatest velocities of wind and the mean velocities, etc, also a diagram showing the relation of the number of bursters in each hour of the day to the barometer curve.

THE PRIZE ESSAY.

In the early days of Australian settlement, when the shores of Port Jackson were occupied by a sparse population and the region beyond was unknown wilderness and desolation, a great part of the Haymarket was occupied by the brickfields from which Brickfield Hill takes its name.

A BRICKFIELDER.

When a "Southerly Burster" struck the infant city its approach was always heralded by a cloud of reddish dust from this locality, and in consequence the phenomenon gained the local name of "brickfielder." The brickfields have long since vanished and with them the name to which they gave rise, but the wind continues to raise clouds of dust as of old under its modern name of "Southerly Burster." A consideration of the earliest reliable records, and a comparison of them with these of later times, appears to prove that the phenomenon itself, as well as its surroundings, has changed.

SIXTY MILES PER HOUR.

Even up to within ten or fifteen years ago the velocity of wind was frequently as high as sixty miles per hour, and occasionally attained the tremendous force of eighty miles—on one memorable occasion it went far beyond this and registered the unprecedented velocity of one hundred and fifty-three miles per hour in a gust.

ONLY FORTY MILES PER HOUR.

Now the southerly burster seldom exceeds fifty miles, and generally ranges between twenty and forty miles per hour. Whether this result arises from the fact that civilisation has raised much brick and mortar to obstruct the atmospheric disturbance, or whether it is that the absorption or radiation of heat is less from the cultivated soil than from the hard, unbroken surface of the pre-cultivation days, is a matter of conjecture only.

BURSTER ANTECEDENTS.

The climatic conditions preceding a southerly buster are, first, a period of high temperature varying from three hours to three or more days, accompanied in the early part of the summer, or towards its close, by wind from the west or north-west, and in the midsummer months, generally from the north-east. In the early morning on the day of a "burst" the sky is white and hazy of aspect. As the hour of the outbreak approaches there begin to accumulate in the south, ball-shaped cirro-cumulus or pilot clouds, and frequently, if electric disturbances are to accompany the squalls, heavy cumuli thunder-clouds rise gradually in the south-west.

THE CLOUD ROLL.

An hour or so before the squall, a heavy cumulus roll appears low down on the southern horizon—the interval between this apparition and the beginning of the gusts depends entirely on the velocity of the wind. Afar off this cloud roll appears most frequently due south, but sometimes south-south-west, or even south-west; it is sharply defined, dark on the edges with lighter shades towards the centre. The roll is from thirty to sixty miles in length. Sometimes it appears singly; on other occasions there are a multitude of these formations heaped one above the other, with light cirrus below. Generally, if the burst is of the first order, it is followed by an overcast sky composed of nimbus from which patchy rain descends. (See *Plates* 1 to 4.)

As the cloud roll approaches it gradually loses its symmetrical aspect, and careful observation reveals a light cirrus fringe rising from underneath it in front, falling over the top, and trailing behind the advancing cloud. Up to this time the wind has been blowing with steadily increasing force from a northerly direction,

INTERVAL OF CALM

but at this stage it drops suddenly and a profound calm prevails, broken only at intervals by a few fitful puffs. This state of things lasts for a varying space; if the southerly arrives during the heat of the day, it endures but a very short time; if at night the calm is of longer duration.

THE BURST.

Meantime the cloud roll is seen rapidly approaching, clouds of dust rise in the southern part of the city four miles away, and gather volume as they come, until they almost hide the city as viewed from Observatory Hill, while from immediately under the roll light clouds rush forward with great velocity only to be thrown back over the top of it as they reach the front, the wind vane on the Time Ball tower flies to the south and the wind reaches us on the ground a moment later, and in a few moments is blowing with the force of a gale.

VERY LITTLE DEW.

It is noteworthy that the night preceding a burster, however clear it may appear, seldom precipitates any measurable quantity of dew. This affect is, no doubt, owing to the excessive dryness of the northerly current which absorbs any moisture obtainable by radiation from the earth, or by condensation from the upper strata.

A BURSTER FOLLOWS A FOGGY MORNING.

But it is a curious circumstance that during a period of hot weather, should a fog exist at daybreak a southerly change is almost certain to follow within twenty-four hours. Such a condition is of rare occurrence and is always preceded by high pressures without energy or grade, and consequently at a time when land and sea breezes prevail.

THUNDER AND RAIN.

A burster rarely brings immediate rain, except when it is accompanied by a thunderstorm. In the event of a three days blow the downpour, if any, seldom takes place until the second day and even then—as is the case with all coastal rains—is heaviest on the promontories, but little falls over the eastern slopes of the mountains and it seldom reaches the highlands. In nearly all cases the coastal region benefits by rain from the southerly half of the following anticyclone.

CONDUCIVE TO DRYNESS.

As noted above, the burster is itself conducive to dryness rather than rain, it being caused, apparently, by the hot, dry conditions prevailing on the plains. The rain by which it is accompanied is caused entirely by electrical disturbances.

A GRADUAL VEERING OF WIND TO SOUTH WITHOUT A BURSTER.

This fact may be demonstrated by studying the progress of a southerly change when it is not accompanied by the "burst." As the advance isobars of the approaching anticyclone reach Central Australia, the northern part remains stationary while the southern expands eastward with rapidity. The barometers rise rapidly in Tasmania and the region of high pressure, then extends itself northwards east of this coast, and thus in part surrounds a pocket of low pressure. This low pressure area is then forced northwards or towards the equator, but not to the eastward, and this temporary stoppage of the usual easterly motion is the precursor of gales, not only in the burster season but at all times in the year. In the course of the further development of this system the low pressure is ultimately forced to north-east off the coast of New South Wales, and the isobars trend from north-west to south-east about the thirty-second or thirty-third parallels of latitude.

STRAIGHTENING OF ISOBARS AND RAIN.

At the same time the rear isobars of this retreating high pressure are becoming flatter or straighter, and causing a divergence of the south-east trades—which prevail almost constantly north of New South Wales—into a north-east wind. The result of this movement is an inflow of humid air which penetrates into the southern part of the continent as far as the South Australian border, and sometimes still further west, that is, six to seven hundred miles.

SOUTHERLY WIND THE PRECIPITATING AGENT.

As the following high pressure approaches, the southerly or surface precipating current in front of it travels eastward, and meeting this humid atmosphere rain *invariably* falls inland, the most copious downpour being along the western slopes of the mountains. Ultimately this approaching anticyclone reaches the eastern sea-board, thereby forcing the low pressure off and up the coast, and by this time the wind which was southerly inland has a tendency to veer to east following the isobars, and develops one of the ordinary forms of easterly weather, which although sometimes dangerous to mariners by reason of its high winds and stormy seas, is by no means so hazardous as the burster. The wind rises gradually, and thus several hours warning is given before the disturbance assumes a serious aspect.

A change of this character rarely makes itself felt beyond the Queensland border. While the fury of the storm is spending itself, the high pressure continues its forward motion south of the storm, the low pressure fills up, and the storm is over.

The weather during the forty-eight hours preceding this change may be, and generally is, locally hot and fine, but the developments in the twenty-four hours immediately antecedent to the outbreak are exceedingly rapid. On the day of such a change the weather reports, even from the inland districts, may tell of fine weather, but by evening the disturbance is felt in the far west, travelling with great speed towards the highlands, and during the night, or by daybreak, it reaches the coast lands.

LOCAL CONDITIONS.

The weather prevailing locally presents exactly the same features as that which heralds the southerly burster, and unless the observer is furnished with complete data, both from South Australia and Queensland, it is practically impossible to differentiate the conditions.

DIFFICULTY IN FORECASTING.

Given these data however, it is possible to foretell accurately a disturbance of the kind above described. So far back as the date of the very first of the series of daily weather charts upon which this essay is mainly founded, the conditions which I have endeavoured to describe have produced the same results without exception. I have thus traced the changes by which the conditions promising a "burster"—*i.e.* an approaching anticyclone with its low pressure Λ—are modified locally so that we have no southerly burster but in its place a south-east gale. There are other though less imprtant factors, tending to rob the southerly wind of its velocity.

RAINFALL IN THE WESTERN COLONIES.

The first and most important of these is rainfall, only in this instance the precipitation takes place further to the westward. The anticyclone appears on the coast of West Australia accompanied by severe gales and rain, which have, in a great measure, wasted themselves before the disturbance reaches Bass Straits, and on the coast of New South Wales only the last dying breath of the atmospheric upheaval is felt. The barometric evolutions may be related as follows:—In the first place the high pressures both to east and west of the Λ depression, are in unseasonable latitudes, the summer tracks lying between 36° and 37° south, and never travelling much below that. At first the adjacent isobars of both high pressures are very full and round; the col between is narrow, subsequently the conditions become intensified, the isobars straighten, and the Λ becomes acute.

EFFECT OF COASTAL FORMATION.

As they advance, the anticyclones gradually become separated by a wider space and isobars of the depression open out and become round instead of pointed, grow more obtuse in fact, and were it not for the mountains of the east coast intensifying these conditions the wind should reach us as westerly instead of southerly.

BURSTERS IN SOUTHERN COLONIES.

While the barometric conditions just referred to are travelling over West and South Australia they produce weather somewhat similar to that experienced here under similar conditions, and hence it is said that the burster is felt in South Australia and Victoria as well as on the eastern coast, but although this is true in a sense, the local conditions on the east coast intensify the characteristics of the "burster" to such a degree that it is very unlike the burster of the southern Colonies, and is in fact quite different.

GEOGRAPHICAL CONDITIONS IN SOUTHERN COLONIES.

In South and Western Australia particularly, the atmospheric waves pass over comparatively level country, and the fact that two high pressures are sometimes so close together may be ascribed to accident. But in New South Wales the geographical aspect is widely different. This province is traversed along the east coast by a mountain range, whose pinnacles have a mean altitude of from three to four thousand feet, and this checks the forward motion of the anticyclone until the moment of the mass of high pressure carries it over, but while the power to do this is accumulating there is formed in the hollow or vast basin, west of the

BURSTERS INTENSIFIED ON THE PLAINS.

highlands, what may be termed a local depression, the sun heating the plains and causing an up draught from the soil, which makes the A assume an intensified character. And it is retained in this situation until the high pressure at its back gathers sufficient force to send it also over the mountains, which it ultimately does suddenly, and the southerly wind relieved of the surface friction of the land comes up over the coastal water with a violent rush or " burst."

DIVERTING THE ANTI-CYCLONES TO SOUTH EAST.

Under the influence of this accumulated motive power which has placed the burster on the coast, the anticyclone is occasionally diverted from its easterly course to a south-east direction, and instances are recorded where, so far as it is possible to trace its movements it has disappeared travelling almost due south.

ANOTHER CONDITION MODIFYING BURSTERS.

There is yet another condition, or set of conditions, which not only modifies the southerly burster but retards it to a considerable extent. As a general rule when there is a Λ depression lying with its centre approximately south of Wentworth, it may be confidently anticipated that the southerly will reach the sea-board after a period varying from twelve to eighteen hours.

EFFECT OF ELECTRICAL CONDITIONS.

But should the front of the Λ develop more than the normal electrical activity, it will inevitably be delayed for another day, and in the meantime the barometric conditions advance but slightly. A southerly of this type is almost invariably only one of a series. The local weather features that herald its advent are the same as those which foretell the coming of the ordinary burster, except that the temperature on the day of thunderstorm is higher than that of the day on which the burster reaches us. The isobaric curves accompanying this particular form of disturbance are notably disorganised and irregular to the east of the Λ, while on the high pressure to the west they remain smooth and even.

TYPES OF LOW PRESSURE.

Southerlies have been observed to result from three distinct types of low pressure. The first of these is the familiar Λ depression, resulting in the true southerly burster. This is the one most commonly experienced, and as a rule, the sharper the ∧ the more sudden is the change.

THE MONSOONAL DEPRESSION.

The second variety is the tropical depression or tongue which may be looked upon as an inverted ∧, which only occurs during the monsoonal season, and even then only on rare occasions; in fact the tongue has never been observed to exist east of the mountains, though west of the range it has been known to reach to the Victorian border. This peculiar type of disturbance is accompanied by an overcast tropical sky, which almost invariably is found preceding the southerly current.

BURSTERS WITH HIGH TEMPERATURE.

The temperature is very high; thunderstorms are prevalent; and the southerly itself is not, generally speaking, very strong. Still it is, perhaps, more beneficial than the first description of southerly, as it is attended by rain in the northern parts of Australia, and west of the highlands in the northern parts of New South Wales.

TROPICAL DEPRESSIONS.

Should the tropical tongue, when it retreats towards the equator travel east, as occasionally happens, the lower part seems to become detached and forms an active rain cyclone on the south-east coast of Queensland. The floods which took place in that locality early in 1892, as well as those on the northern coast of New South Wales, may be ascribed in part to this cause. This opinion, however, it is necessary to mention, rests upon few data and must be taken subject to correction.

The apparent parting of the tropical tongue to which I have alluded is by no means the invariable, or even the most common form of development. The low pressure always exists in a fluctuating state, north of this continent, the isobars sagging down between the high pressures as they pass along, and drawing up again as the high pressures pass under them. In this way they act upon the low pressure to much the same effect as a succession of ocean waves might do, while passing under a long, pliable, floating body set at right angles to them.

BURSTERS FROM SECONDARIES.

The third and last modification results from a secondary. Of this particular variety very few have been recorded on the weather charts. They develop so suddenly that unless they are actually in a state of evolution at the hour when observations are taken (9 a.m.), the meteorologist has no indication of their proximity. They develop on the south coast of New South Wales through the formation of a "kink" in the outlying isobars of the retreating high pressure. Prior to such a development the barometric conditions are neutral or dormant—in other words, the existing high and low pressures have little grade, and are only relatively high and low. Southerlies caused in this way partake of the well known characteristics of secondaries, wherever they may occur—that is, violent and of small compass, most severe in the extreme south east, and seldom affecting the coastal districts north of Cape St. George. The depression which follows the retreating high pressure brings on the following day, a southerly of the ordinary character, the strength of which is regulated according to the intensification or otherwise which has taken place during the interval in the adjacent high and low pressures.

BURSTERS NOT CONDUCIVE TO RAIN.

The next point to be considered is the relation of bursters to rainfall. The burster, on the whole, must be regarded as unfavourable to rain, for though its advent may benefit the coastal areas to the extent of a few showers it serves as an indication that the country west of the ranges is at least in a temporarily dry condition. A succession of bursters especially denotes intermittent rain in the interior. As a proof of this, and also as an evidence of the comparative powerlessness of merely local weather conditions to directly cause, or even to intensify, the burster, it is only necessary to point out its frequent occurrence under circumstances apparently most unfavourable for its development.

SOUTHERLY BURSTERS.   27

ARRIVES WHEN TEMPERATURES ARE LOW.

It often arrives during a period of low temperature at Sydney, and also when rain has actually been falling—on some occasions when it has been falling heavily. This statement is, of course, not intended to imply that the southerly may not be modified by the existence of such conditions upon the immediate scene of its action. Still, despite any modification which may take place, the wind attains a by no means inconsiderable velocity, some storms recording as high a rate as forty miles per hour. Moreover the rain generally ceases and a period of fine weather follows, proving

RAINFALL FROM NORTHERLY WIND

conclusively that, in these particular instances, at all events, the rainfall which takes place on the arrival of the burster is the precipitation from the moist northerly current, and not from the southerly one. Bursters have also been recorded when the local maximum thermometer has registered little over 74° in a midsummer month, and effecting a diurnal range of only four degrees. At such times the temperatures in the interior are always high, especially to the north-west of New South Wales and in southwest Queensland.

BURSTERS DESIRABLE.

Though the individuality of the burster is, in itself, opposed to rainfall, yet the frequent recurrence of this phenomenon is to be desired to cool the atmosphere, and its frequency indicates that the cyclonic systems in the north-east, and the anticyclonic systems in the south have less than the average energy; also in the early part of the year it denotes the failure of the monsoonal rains.

SEVERE DROUGHT COINCIDENT WITH THEIR ABSENCE.

For instance, their absence was very marked during the severe drought in the summer of 1875-76. In seasons when southerlies are rare the loss of the interval of cool weather adds to the severity of the summer.

MONOTONOUS WEATHER.

During such droughty periods as that just referred to, the barometrical changes over the whole of Australia are very slight. The temperatures, however, are frequently very high inland, but the diurnal ranges are not great. Locally, in Sydney, during the mid-summer months, the maximum shade temperature has been known to range between 80° and 86° without hot nights, for twelve consecutive days, with persistent north-east winds. Experiences of this nature are monotonous for the pastoralist and agriculturist, and equally so for the meteorologist, presenting as they do, no interesting features for observation, and no change of moment, either in weather or temperature can be anticipated until the barometer shows renewed activity.

CHANGES OF TEMPERATURE.

Generally the greatest diurnal range of the temperature resulting from bursters occurs in October. The mean minimum temperature for this month is 55·1° so that when the maximum reaches from 85° to 90° and a southerly takes place, the temperature in some cases drops as much as 30° to 35° This range is seldom reached in the hotter months, in fact it is only attained when a maximum of about 100° is recorded—an event of very rare occurrence.

A BURSTER OF THE POPULAR TYPE.

An anticyclone of good energy, and one from which a popular type of southerly burster lasting three days results, has a latitudinal axis of approximately two thousand four hundred miles and as it moves at the rate of four hundred miles per diem,[1] it follows that if a vertical line is drawn in front of it, an observer stationed in this line will record three days of southerly weather in the front half of it and three days of northerly winds in the rear, due to the normal circulation about an anticyclone. This is the actual experience ; provided no meteorological agencies affect or modify its symmetrically oval form during its passage over the Australian

---

[1] See Moving Anticyclones, p. 3.

continent, steady breezes throughout its circulating areas are the rule during the six days it occupies in passing over the comparatively low lands of Australia, as it travels from west to east.

ISOBARS FLATTENED BY MOUNTAINS.

But when the anticyclone reaches the mountains of New South Wales, the symmetry is disturbed by a flattening of the front isobars against the mountains, which concurrently show a tendency to spread northwards, aided probably, by the inclination of the mountain chain towards the north east.

INTENSIFICATION OF THE LOW PRESSURE BY MOUNTAINS.

While this change is taking place in the high pressure, the isobars of the Λ depression in front of it are undergoing a similar process. But as this depression is a much smaller system, the front isobars are being squeezed against the highlands, while those at the rear are also being compressed by the advancing high pressure. The result is an acceleration of force in the northerly and southerly winds which enclose the Λ depression. When the squeezing process has amassed sufficient energy to overcome the obstruction, the gradients of the high pressure, in their efforts to regain their normal condition, expand with great rapidity on the eastern slopes and send the Λ depression at a great rate to the south-east.

For the previous twenty-four hours the northerly currents have been maintained by the Λ depression, but when once the mountains are passed the anti-cyclone takes up the running of the southerly one. As a proof of this—speaking particularly of the central and northern parts of the coast—the southerlies have nearly always been found to circulate along the line of the seaboard in the anticyclone isobars. After an anticyclone of this character has entered by a three days' march into Australia, it is very rarely indeed that its progress is checked ; but occasionally, as the Λ depression still travels on, it seems to drag the southern side of the high pressure after it. At the same time, while the body of the high pressure remains over Australia, the one preceding continues on its way towards New Zealand. This results in an attenuated state of atmosphere on the northern coast of New South Wales and on the shores of Southern Queensland.

TROPICAL CYCLONES.

If it so happens that one of the north-east cyclones is on a visit to this region at the same time it is diverted from its customary course— the track of these storms being generally to the eastward of our coast—and avails itself of this partial void. Its irruption, acting with the already existing a depression, seems to result in that peculiar class of storm, of which the Dandenong gale is a notable example.

Concerning the intensity of this particular storm a few facts concerning another of a similar character may not be considered out of place. A study of its leading characteristics may, in some future time assist the meteorologist in recognising the early signs and portents of another such calamity.

A RECENT GALE OF THE DANDENONG TYPE.

The second gale referred to occurred on the 23rd of September, 1892—sixteen years after the famous " Dandenong " storm. On this occasion the wind attained a velocity of one hundred and twenty miles an hour. The series of anticyclones in August and those in the early days of September were of considerable energy, and in each case the high pressure to the west, which, from its unusually energetic state, doubtless aggravated the violence of the storm, had completely disappeared on the following days. All that remained were horizontal isobars of a relatively low pressure in the southern ocean. Particulars of the "Dandenong" gale may be found in Mr. Russell's paper on "Storms on the New South Wales Coast," read before the Royal Society of New South Wales on 7th August, 1878. (See following diagram, page 32.)

BAROMETERS FALL AFTER WIND VEERS TO SOUTH.

These are two noteworthy instances in which the barometers, after a southerly, have been observed to fall instead of rise, with active high pressure in the rear at the commencement of the blow. The gale of 1892 was not a burster, but a gradual change to south which took place some thirty hours previously. In this instance the Λ depression, which seemed to act in concert with the tropical cyclone, appears to have backed from a situation midway between Australia and New Zealand.

CONDITIONS WHICH REGULATE THE NUMBER OF BURSTERS.

Before proceeding further, it may be advisable to offer a suggestion as to the causes which seem to the writer to justify the two apparently anomalous statements made in an earlier part of this essay: (1) that bursters are less frequent and also less violent than usual, during those seasons when the interior of the country is suffering from drought; and (2) that they are also less frequent and of less than their ordinary violence, when the same region is visited by persistent rains with overcast skies. In other words, seasons of drought and desolation and seasons of deluge are both inimical to the existence of bursters.

The following theory is tentatively submitted, to account for the greater prevalence of bursters during what may be called a moderately dry season. This term is used here to indicate, not so much a season in which the rainfall does not exceed the average, as one in which the number of rainy days is a shade below the average. It will be remembered that strong southerlies have been said to result from energetic anticyclones; also that energetic anticyclones bring with them possibilities of great extremes in our summer temperatures. But since in summer their sphere of action is in high latitudes the weather of the northern parts of Australia is controlled chiefly by tropical depressions, and consequently great heat prevails there as a rule. Now if, after a period of hot weather, rain were suddenly precipitated on this northern area, the consequence would be a great uprush of air with the vapour produced by the rain falling on heated ground, and a consequent inrush of air from southern areas.

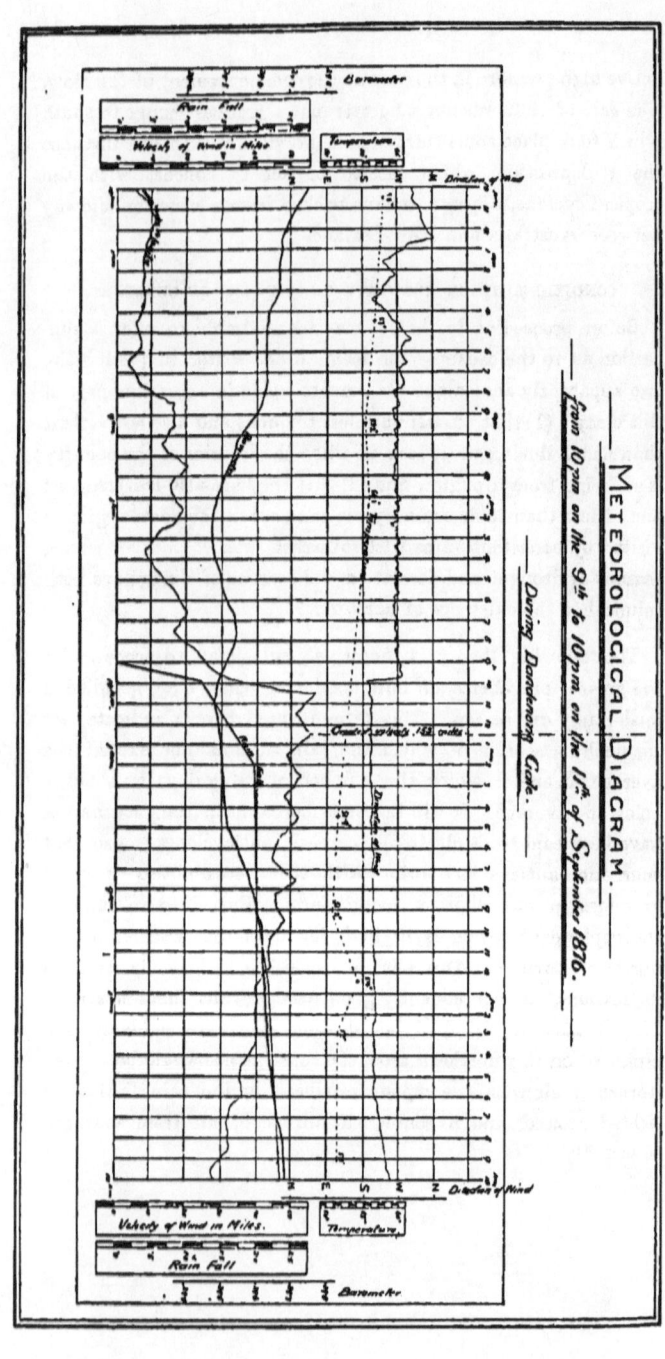

SOUTHERLY BURSTERS. 33

RECURRENCE AND DURATION OF BURSTERS.

Bursters are not only uncertain in their duration, but the periods of their recurrence are extremely erratic. Two have been frequently known to take place within twenty-four hours; between two others an interval of a month has been known to elapse. The shortest one recorded extended over a period of three hours; the longest covered the space of ten days. These figures of course, embrace all the variations of the wind during the continuance of the storm.

SHORT-LIVED BURSTERS.

The short lived bursters are generally experienced during a seasonal prevalence of north-east winds; those of longer duration are usually met with during a southerly prevalence.

PARTIAL TO COAST OF NEW SOUTH WALES.

Not only during the summer, but in all seasons of the year, the southerly has a remarkable partiality for the sea-board of New South Wales.

OVER STEP ISOBARS.

So much is this the case in the hot months that the approaching currents from the south have been observed on the coast when the isobars indicated northerly winds.

BLOW CONTRARY TO ISOBARS.

Sometimes, too, for several consecutive days, the wind has blown with appreciable velocity from this quarter when the isobaric lines—from their relation to the centre of the anticyclone—should be producing northerly winds. This latter pecularity is most noticeable in winter months and with necessarily small gradients.

THE AREA OVER WHICH THE BURSTER EXTENDS.

The land area over which the burster exercises its influence may be generally described as including all the country east of the mountains in New South Wales, from the extreme south-east point of the colony to a little above Port Macquarie on the north coast. It occasionally oversteps the northern limit, but when it

C

does so it is usually the result of a cyclonic disturbance on the north coast, such as existed, for example, in the "Dandenong" burster, the effects of which were felt considerably north of Brisbane. The northern boundary of the burster may be defined, as limited by the south-east trade winds, which blow almost incessantly north from about 30° S. latitude. Other areas of actual experiences are given on page 40.

CHANGE OF TEMPERATURE.

Bursters always result in a diurnal drop in temperature, the said drop ranging between 37·5° and 4·2° with an average fall of 18·1°. The greatest diminution of temperature takes place during the first hour, and the fall is most sudden when the burst comes at midday. The steepest drop on record was that registered at 1·50 p.m. on the 30th December, 1891. The maximum then read 97·5°; five minutes after the change to south it read 80·3°, being a difference of 17·2°; two minutes later there was a fall to 79·7° or a further difference of 0·6°. The thermometer remained at this reading until after two o'clock, but at 2·15 it descended to 75°, being a drop of 4·7° more and a total drop of 22·5° in twenty-five minutes. The diurnal range in this instance was 23·8°.

VELOCITY OF WIND IN THE BURSTER.

It occasionally, though rarely, happens that the greatest velocity of a burster is reached at the change or during the first hour thereafter. In most cases, however, the maximum force is attained about twelve hours afterwards. As the majority of bursters occur between the hours of seven and twelve p.m. (see Table I.), it therefore follows that the wind is usually strongest between seven and twelve a.m. on the following day.

UNSEASONABLE WINDS.

In these instances where the greatest velocity was reached immediately on the arrival of the burster, the winds preceding were generally of an unseasonable character—that is to say north-westerly in the summer and north-easterly in the spring and autumn months. Under these circumstances the burster, though strong, was generally of brief duration.

THE PAMPERO.

The opinion has long been held that there is a close analogy between the southerly burster of Australia and the pampero of South America. The writer therefore sent a detailed description of the burster to a friend formerly resident in Brazil, and requested him to note the points of similarity between it and the pampero. The following is a copy of his reply which is based, as will be seen, partly on personal observation and partly on information gathered by him from sources on which he places reliance :—

"In answer to your letter of 14th of April, I am afraid I can afford you very little information on the subject of the South American pamperos from personal observation. They never reached, on the coast, as far north as Rio de Janeiro which was my usual place of residence, but I twice experienced their violence when on my exploring expedition across Brazil in 1872-3. The first time was in September, 1872, when we were camped on the banks of the Paraguay River as far north as 20° south. On the particular occasion we were awakened in the night by a roaring sound which rapidly drew nearer and burst upon us overturning tents and everything else that offered much resistance without being stable. My hammock was slung to a branch of a big tree, which was torn off, but did me no damage as I had turned out to save the tents. It had all blown over by the morning. On the second occasion, in October, 1873, I was surveying the Alto Parana River, about 21° south, when my Indians noticed the clouds gathering in south, and made at once for the lee of the islands in their canoes. The river at this part is due north and south for many miles, and from one to four miles wide, and as the wind came from due south it beat up a great sea, the waves being quite three feet from crest to bottom. I sailed up before it, and was able to go up rapids that, on ordinary occasions could only be negotiated by poling and ropes. It did not last long, and on both these occasions was, I imagine, a more violent effort than usual of a pampero which found its way so far north owing to the immense width and openness of the river Plate Valley, of which river both the Paraguay and Parana are confluents. The pamperos are mostly felt at the mouth of the River Plate, where by their violence they often cause considerable damage to shipping. If they are not recorded further south, it is because there are no ports on the east coast nor inhabitants or towns where their recurrence could be noted.

"In the southern States of Brazil, they are frequently felt, and a friend of mine who has just arrived from Rio Grande du Sul, the most southerly state of Brazil, tells me they occur mostly in the winter months (my friend was asked to verify this statement and his reply will be found further on.—H.A.H.), from April to November. He says they are preceded always by three days rain, which, when over, is immediately followed by the pampero, coming from what he called the south-south-west or nearly due south. They last then for nearly three days and blow with great violence and low temperature. My informant goes on to say, from his own experience, that the pampero does not get so far north on the coast, although, as I have told you, in the interior it does seem to penetrate. There are high ranges of mountains south of Rio de Janeiro which would, I have no doubt, have the effect of breaking up and deflecting any part of the pampero which might go there."

The friend's account is as follows :—" Pamperos proper occur in the winter, say from the end of May till October, and generally last three days. Occasionally in the summer time we have, in Rio Grande, smart breezes from the south-west after rain, but they do not last long, and although they come from the same quarter and cool the atmosphere wonderfully, are not called pamperos. While the pampero is blowing the sky is beautifully clear and cloudless. Occasionally in the summer in Rio Grande when the weather is very sultry, it breaks by a squall suddenly setting up from the west by south-west, and though the sky was clear ten minutes before, you can see the cloud roll forming in the west as the cool breeze advances, and down the rain comes in torrents, and in a few hours all, wind and rain, is over. Thunder may or may not accompany the rain, which almost always precedes (at least in Rio Grande) pamperos. I have seen pamperos as strong after rain, when there was no thunder accompanying, as when there was. Our thunder, too, is generally, in the summer and rarely in winter. I must add that Rio Grande being further north than the Plate, we do not get the pamperos with the full force experienced in the Plate region. With us they are a steady continuous blow for three days, varying little in force till the third day, when they are felt to be gradually declining. They always blow from the same quarter, the south-west, and are cold and dry."

The most interesting fact evidenced in this description of the pampero is that it always follows rain. This would seem to imply that the evaporation arising from the plains is one of the immediate causes of its existence. If this is the case it lends support to the theory, hereinbefore submitted, that the vapour arising after rain from the vast heated areas of the Australian interior is at least one of the agencies from which the burster has its origin. The difference between the two sets of circumstances is that, in South America, the cause and effect act on the same region, and therefore one follows the other with no appreciable interval, while in Australia the seat of the effect is removed by something like one thousand miles from that of the cause, and consequently the connection is more difficult to trace.

Winter northers of Texas are a somewhat similar experience. The following is a short extract:—"The northers prevail from November to March, and commence with thermometer at 80° or 85°. A calm ensues on the coast; black clouds roll up from the north; the wind is heard several minutes before it is felt; the thermometers begin to fall; the cold northers burst upon the people bringing the thermometers down to 28° and sometimes even to 25°, men and cattle being killed from the severe cold." This is the only description I have come across of similar changes in the northern hemisphere. [1]

---

[1] Physical Geography of the Sea by Maury, pp. 93-94.

SIMILARITY OF WEATHER ON EAST AUSTRALIA AND SOUTH AMERICA.

Before finally leaving the subject of the Pampero, I may mention that many old sailing ship captains recognise numerous points of similarity between the southerly bursts upon the east coast of South America and those upon the east coast of Australia, and make the same preparations to meet either case.

ADMIRAL FITZROY AND OTHERS ON THE PAMPERO AND BRICKFIELDER.

Perhaps the earliest reference to the pampero appears in the "Weather Book" of Admiral Fitzroy, pp. 150 and 151. In chapter xxi., he gives thrilling accounts of two occasions when his vessel was struck by it and nearly foundered. The same author, in the same book, speaks of the brickfielder (southerly burster) of New South Wales, page 263. A more exhaustive account of the pampero in its strictly scientific aspect, appears in "Weather," by the Hon. Ralph Abercromby, page 263, and a description will be found in the Scottish Meteorological Journal, Vol. v., No. lx., p. 335.

A DESCRIPTION OF A BURSTER WRITTEN EIGHTEEN YEARS SINCE.

The following description of a southerly burster is taken from "The Climate of New South Wales," by Mr. H. C. Russell, B.A., The passage in question contains the best description of it that I have met with, and it is the more interesting as being one of the first accounts published by any scientific observer of this interesting phenomenon. It is as follows:—"If in fine north-east hot weather the barometer falls fast in the forenoon, a southerly wind (burster) may be expected before night; if the day is very hot the change will come sooner; and if the barometer is falling very fast and clouds be seen in the west, a thunderstorm may be expected in the afternoon. Sometimes the thunderstorm bursts first, and the wind sets in from south afterwards; if only the storm comes it will probably be hot again next day.

"Southerly bursters are generally to be expected from November to the end of February; they are always attended with strong electrical excitement, a stream of sparks being sometimes produced

SOUTHERLY BURSTERS. 39

for an hour at the electrometer at the ends of the exploring wire. The approach of the true burster is indicated by a peculiar roll of clouds, which when once seen cannot be mistaken; it is just above the south horizon, and extends on either side of it 15° or 20°, and looks as if a thin sheet of cloud were being rolled up like a scroll by the advancing wind.

"Clouds of dust, which penetrate everywhere, announce the arrival of the wind, scud flies by overhead with great rapidity, being sometimes less than two thousand feet high; rain may follow, but if so, thunder and lightning come first. The velocity of wind is, in most cases, greatest within the first two hours, and varies from thirty to seventy miles per hour, but is usually from fifty to sixty and the rate of progress along the coast about forty miles per hour. The change of wind is sometimes very sudden; it may be fresh north-east, and in ten minutes a gale from south, hence vessels not on the lookout are sometimes caught unprepared, and suffer accordingly. When the wind is light these storms are often carried to sea by the general easterly motion of the atmosphere, and may be seen passing by, the peculiar clouds indicating unmistakably their position."

APPARENT DOUBLE BURSTERS.

The following is a detailed account of a double burster, or of two distinct simultaneous bursts on different parts of the coast. It will be convenient to commence with a table, showing the hours at which the burst reached various points on the coast-line, with relative rates of travel from place to place:—

10th October, 1893.

Stations.	Distances.	Hour.	Rate of Travel. Interval. h. m.	Miles.	
Eden	—	2 a.m.	—	—	
Moruya	79	8·30	6 30	12·2	
Jervis Bay	47	10·30	2 6	22·4	Wind velocity at Sydney 35 miles per hour.
Wollongong	41	11·45	1 9	32·8	
Sydney	41	1 p.m.	1 15	32·8	
Newcastle	62	3·40	2 40	23·3	
		Mean rate of travel		25·3	
Port Macquarie	110	2 p.m.	—	—	
Clarence	122	5·30	3 30	34·9	

From this table it will be seen that the wind changed to south at Eden, the southern point of the coast at 2 a.m., and made its way gradually up the coast, reaching Sydney at 1 p.m. and Newcastle at 3·40 p.m. in the mean time that is at 2 p.m. Port Macquarie, one hundred and ten miles north of Newcastle, reports the change at 2 p.m., or one hour and forty minutes before it got to Newcastle, or to put it another way, when one burster reached a point on the coast twenty miles north of Sydney, another made itself felt at Port Macquarie, one hundred and ten miles farther north, it appears then that there were two bursters separated by one hundred and ninety miles and simultaneously making their way along the coast. It is probable, however, that the southerly burster is a *line storm*, and that the change of wind occurs along a certain isobar, and if this be the true explanation it is entirely in accordance with what has already (page 29) been shown as to the trend of the isobars in certain cases. When the approaching anticyclone flattens its isobars against our dividing range, and then taking advantage of the easier western slope in the northern districts, and the lower altitude of the mountains generally, to the west of Port Macquarie, protrudes some of the isobars over the mountains, so that they come to the eastward at that point like a nose or easterly extension of the isobars, while the southern half of the anti-cyclone is retarded. This feature was noticeable upon the occasion in question, and, as is invariably the case under such circumstances, thunderstorms were prevalent. These, I think, may be safely accepted as the causes which give rise to the phenomenon of two simultaneous, yet entirely independent, bursters.

In speaking of the areas over which the burster exercises its coastal influence, mention has been made of the fact that it is rarely met with north of Port Macquarie. The present case was not, in reality, any exception to the general rule; the burster did not reach Clarence Heads as such, and the change there was simply a veering of the wind from north-east through west to south-east, accompanied by a thunderstorm and hail. The downfall of hail was heavy and some of the stones were very large. This storm lasted on the north coast until 7 p.m., and was also very severe in the northern highlands. It was accompanied by a rainfall of from one to three inches.

DESCRIPTION OF THE DOUBLE BURSTER.

The notes taken of this burster—the second of the season—are as follows :—During the 7th, 8th and 9th October the barometers all over the continent showed little or no grade. The centres of the high pressures were situated, one over the Tasman Sea, and the other off the west coast of West Australia, with a shallow trough of low pressure between. On the 10th the western high pressure had intensified and had made much progress, compressing the low pressure into a sharp Λ, with its axis lying in a north-west and south-east direction, or from Bourke to a little east of Gabo. Strong north to north-west winds were experienced on the western borders of Queensland and in New South Wales north-east, while exceptionally hot weather was reported from the Queensland inland stations. In Sydney, for three days previous to the arrival of the burster, the sky was hazy and almost tropical in its aspect. The barometer at Sydney fell three-tenths during the twenty-four hours immediately preceding the burst, and rose rather sharply after it had passed, the lowest point making half-an-inch in twenty-four hours. The temperature was moderate immediately before the change, the thermometer stood at 75°, and at 2 p.m., rather less than one hour after, it had fallen to 65°. A roll of cumulus cloud of a rather undefined character was first seen at 12·30 p.m., with a line of ragged cirrus beneath. The latter, as the storm advanced, rose in front of and obscured the cumulus. Five minutes later, cirrus were moving horizontally, vertically, and in every other direction from a point in the cloud lying due south. The change of wind at Sydney came at 1 p.m., and was attended by much dust, and some rain which was entirely owing to electrical influences, it was generally light from the western slopes to the coast.

Passing from this record of the double burster, it is now necessary to consider various facts relating to the time taken by bursters in travelling along the coast.

RATE OF TRANSLATION OF BURSTERS.

From the preceding remarks and the figures quoted in the table page 39, as well as other figures appended to page 43, it will appear that no definite relation can be traced between the rate of translation and velocity of the wind in bursters. As already demonstrated, it is quite possible for two bursts to occur on different parts of the coast at the same time, and it is also possible for a

BURST SIMULTANEOUS OVER A WIDE AREA.

burst to be felt at the same moment over an extensive area. (See table 28th November, page 43. Since there is no visible connection between the velocity of the wind and the ratio of translation of the burst itself, it may throw light on the matter if we look for some explanation of the fact that the southerly change is generally first experienced on the south coast of the Colony.

BURSTERS FELT FIRST SOUTH OF SYDNEY.

The most probable explanation is as follows :—The high pressure following the Λ depression in many cases moves faster over Victoria than it does over this Colony, and thus forces the lower part of the Λ to the east making the axis of it more or less towards south-east and north-west. Were it not for this swinging of the Λ depression, the wind, from its natural inclination to the centre of a low pressure, would be south-west instead of due south.

CONDITIONS GOVERNING THE RATE OF TRANSLATION.

The southerly isobars of the Λ depression usually reach the south coast first, hence it follows that the burster touches the coastal stations in rotation, beginning at the south. The rate at which it travels in its northerly course is decided, first by the inclination of the axis of the Λ to the trend of the coast, and secondly, by the then prevailing rate of the general atmospheric motion to the eastward.

SOUTHERLY BURSTERS.

DIFFICULTY IN FORECASTING TIME OF BURSTER'S ARRIVAL.

These few remarks will show how many and how serious are the difficulties to be contended with in predicting the moment at which a burst will arrive at any given point for up to the present moment it has been found impossible to decide how far the axis of the ʌ may deviate from its vertical position before reaching the coast, or whether the rate of atmospheric motion may diminish, maintain or increase.

TABULAR STATEMENT OF THE PROGRESS OF SOUTHERLY BURSTERS UP THE COAST.

Stations.	Time.	Distances. Miles.	Interval. h. m.	Translatn. Rate ⅌ hr. Miles.	
13th November, 1893.					
Moruya	4·45 p.m.				Velocity of Wind at Sydney 36 m. in short gusts 42 miles.
Jervis Bay	5·30 p.m.	47	0 45	62·7	
Wollongong	...				
Sydney	7 p.m.	88	1 30	58·7	
Newcastle	11 a.m.	62			
16th October, 1893.					
Jervis Bay	2·30 p.m.				Velocity of wind at Sydney 20 miles.
Sydney	9·50 p.m.	88	7 20	12	
Pt. Macquarie (Oct. 17)	8·30 a.m.	172	19 20	9	
28th November, 1893.					
Moruya	8·10 a.m.			,,	Velocity of wind at Sydney 48 miles
Jervis Bay	7·30 p.m.	47		,,	
Wollongong	7 p.m.	41		,,	
Sydney	7 p.m.	41		,,	
Newcastle	10 p.m.	62	3 0	20·7	
Port Macquarie	2 a.m.	110	4 0	23·3	
30th November, 1893.					
Moruya	2·40 p.m.				
Jervis Bay	4·15 p.m.	47	1 35	29·7	Velocity of wind at Sydney 26 miles.
Wollongong	6·15 p.m.	41	2 0	20·5	
Sydney	7·15 p.m.	41	1 0	41·0	
Newcastle	9·30 p.m.	62	2 15	27·5	
6th December, 1893.					
Moruya	1 p.m.				
Jervis Bay	3·30 p.m.	79	2 30	31·6	Velocity of wind at Sydney 40 miles.
Sydney	6 p.m.	82	2 30	32·9	
Newcastle	10 p.m.	62	4 0	15·5	
Port Macquarie	2·30 a.m.	110	4 30	24·4	
12th December, 1893.					
Jervis Bay	11 p.m.				Velocity of wind at Sydney 25 miles.
Sydney	4·30 a.m.	82	5 30	15	
15th February, 1894.					
Moruya	9 a.m.				Velocity of wind at Sydney 21 miles.
Cape George	1 p.m.	79	4 0	19·8	
Sydney	5·55 p.m.	82	4 55	16·7	

CHARACTER OF WINDS PRECEDING A BURSTER.

When the wind preceding a burster blows from some point between north-west and west, it is always drier and hotter than when it proceeds from between north-east and east. The latter, however, produces peculiarly uncomfortable and relaxing effects, such as might be experienced by one who went into a steambath. The north-westerly wind on the other hand is of a dry and parching character, but apart from its occasionally irritating effect upon the nostrils, and the stinging sensation which it sometimes causes in the eyes, it is endurable, and is enjoyed by many persons and much preferable to the sweltering north-easter.

As would naturally be imagined, when the northerly wind is strong the southerly by which it is succeeded is strong likewise, being effects arising from the same cause. This intimate relation between the force of the two currents is especially noticeable should the burst occur while the northerly wind is in full force— in other words, should it occur in the daytime, for northerly winds, and particularly those from the north-east, moderate considerably by sunset.

RELATION OF VELOCITY OF NORTH AND SOUTH WINDS.

But while it may be said that, as an almost invariable rule, strong southerlies follow strong northerlies, the mean velocity of the former is generally less than that of the latter, by about ten per cent.

WITH STRONG WINDS THE BAROMETER RISES SLOWLY.

And it may be noted, as a peculiar circumstance in a gale with such antecedents, the barometer rises very slightly after the first short and rapid rise, the accumulation of pressure in the west, no matter how steep the gradient is, seems to be used up in maintaining the gale on the coast, and along the coast the barometers rise little or nothing. There are remarkable records of cases in which steady high barometic readings have been the precursors of prolonged southerly gales upon the coast, even the diurnal fluctuations being totally ignored. On the other hand cases are on record in which at Sydney the barometer has fallen 0·5 in. in twenty-four hours, and in Tasmania as much as 0·8 in., these differences expressing the ratio of the changes in the same burster at the two places, due to latitude.

DESCRIPTIONS OF PARTICULAR BURSTERS.

The past twelve months have, unfortunately, been a most unsuitable season for the study of the subject with which this essay is concerned, and it has been especially unsuitable for the observation of cloud movements. The bursters during the past year have not only been few in number, forty per cent. below the average, but also unusually mild in character, and hence it has been impossible to find a burster of strongly marked character for study, of the comparatively small number however, the best have been chosen.

The first of these occurred on February 15, 1894, it has been selected mainly because the clouds admitted of better notes than others, and because I am able, through the courtesy of Mr. Russell to submit photographs of it from the Observatory records. At 9 a.m. on the 14th an anticyclone was over Perth in West Australia, another of rather more decided character was over the Tasman Sea, and between these there was a low pressure wanting in intensity; winds were light and without character, and the weather generally hot and unpleasant, but fine, except some passing showers on the coast of New South Wales. Fresh northerly winds were blowing in Bass's Straits. Diagram No. 1 shows the instrumental conditions of this change from 7 p.m. of the 14th to 7 p.m. of the 16th. It will be observed that the barometer began to fall steadily at 10 p.m. of the 14th, and reached the lowest point at 2 p.m. of the 15th, two and a-half hours before the burster arrived, at the time of lowest barometer the wind was from east, and attained its greatest velocity for the day, seventeen miles per hour, at 3·30 p.m. Temperature was highest 79°, between 1 and 2 p.m. of the 15th, and the coming of the southerly at a quarter to six p.m. made but little difference; light rain

46 AUSTRALIAN WEATHER.

began to fall two hours after mimimum temperature was reached, and became gradually heavier from between 2 and 3 p.m. of the 16th, at which time the wind veered a little to west of south with a considerable increase of velocity.

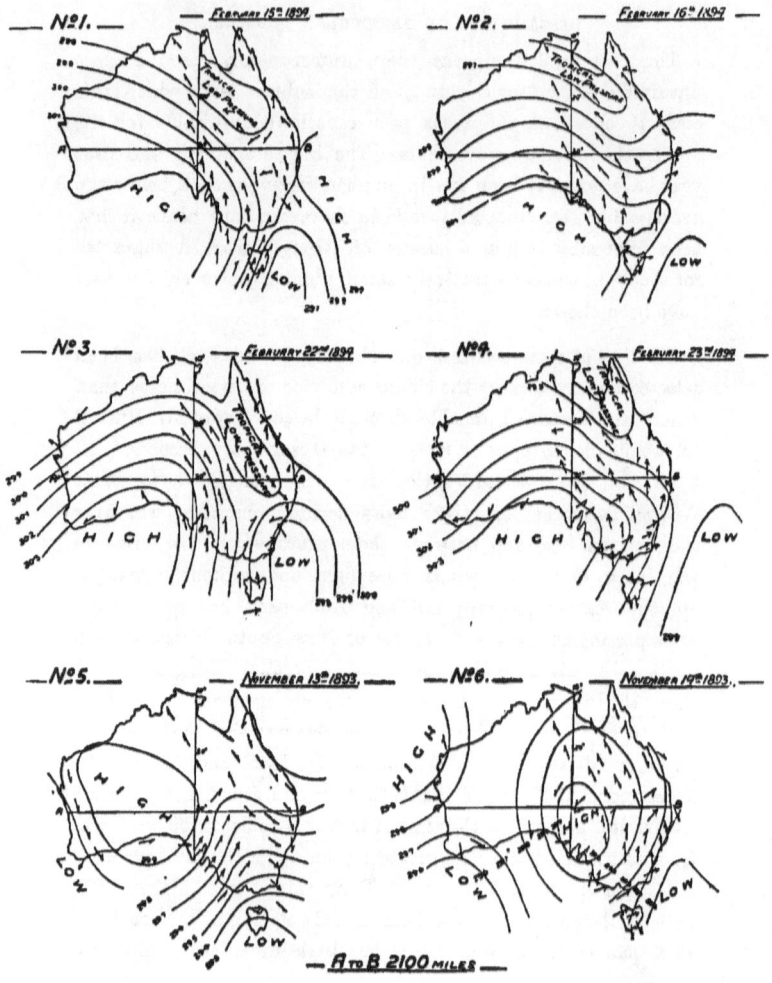

On the morning of the 15th February weather chart No. 1 gives the barometic conditions over Australia and New Zealand, conditions which had materially hardened since the 14th, winds generally were fresher and conforming to the isobars, and weather fine and warm in the eastern colonies. Up to noon there were light cirro-cumulus clouds in the south-west, at 1 p.m. they had become much more dense and somewhat thundery looking, and seemed to be working round the horizon to the north, with one remarkable mass of cumulus. The following diagram of this date shows the instrumental changes during this southerly.

At 2 p.m. south-west clouds slightly advanced, at the same time those low down obscured by haze or dust : up to 2 p.m. north-east, east, and south-east horizons beautifully clear. 2·40 p.m., clouds still working further round to due north, with a few light cirrus forming and evaporating east of this point, and a general drift to the east, then thunder cumulus worked up from south-west very slowly, the advanced cirrus reaching Sydney at 3·30 p.m.; no trace of cloud observable with surface wind up to this time ; a roll of cloud appeared at 3·15 p.m. above south horizon (See *Plate* 1, taken at 3·43 p.m.), which seemed to be made up of a band of stratus surmounted by cumulus 30° in length ; at 4 p.m. the roll merged into the general clouds, which were then fringed overhead and stratified on the southern boundary, beyond which the sky was clear for a space of about 10°, at the same moment an uneven roll of heavy cumulus began to rise above the horizon ; as it lifted, the sky still further beyond was clear. Up to this hour, 4 p.m., only one strata of cloud visible, and that moving from south-west ; a very dense and extensive cloud of smoke to south-west, which worked backwards and forwards between north-west and south-east on the horizon. At 4·20 p.m. the eastern point of the roll was immediately over Botany Bay, the western limit extending indefinitely to the west. (See the details in *Plate* 2, taken at this time.) Small shower at 5·15 p.m. The cloud roll then seemed to melt away like the earlier one, and the upper clouds still moving from south-west became very wild looking,

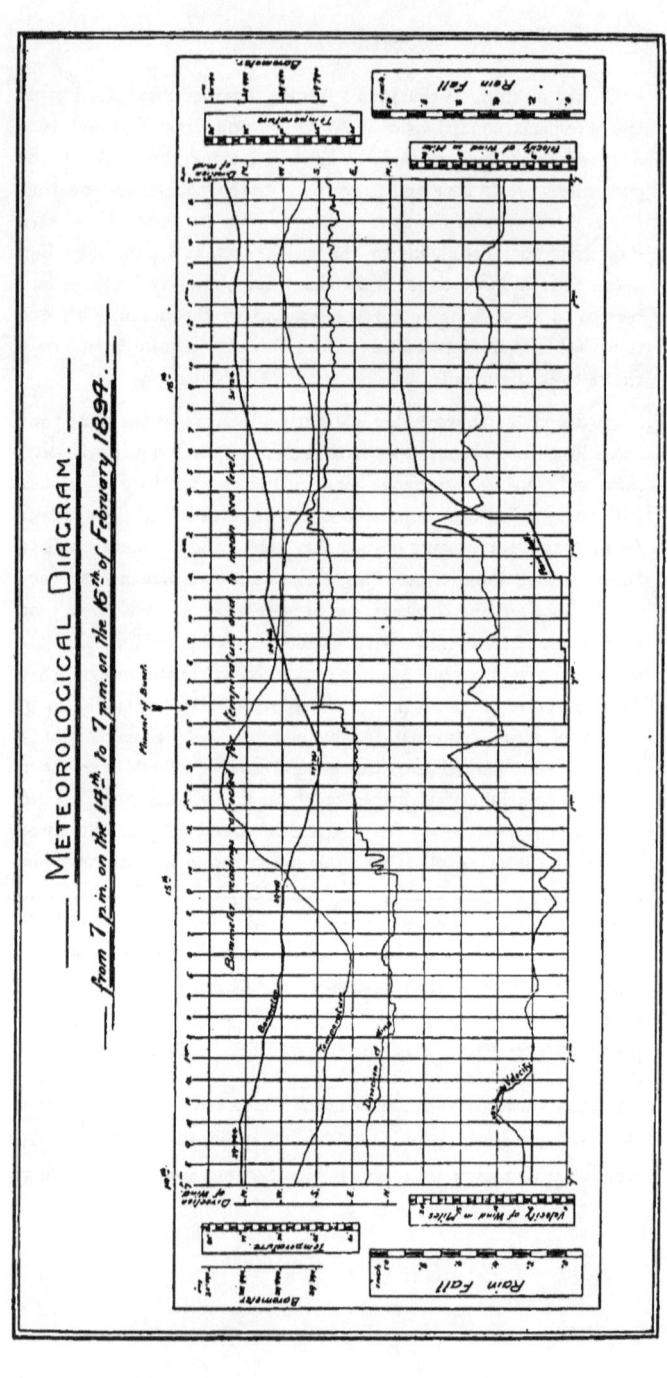

and *Plate* 3 was taken at 5·50 p.m., showing only a trace of the cloud roll and the disturbed looking upper clouds. Southerly burster arrived at 5·55 p.m., but depth of the current very shallow, for the clouds were still maintaining their course from the southwest. The cumulus observed earlier in the evening, at 6 p.m. remained stationary until 7·30 p.m., with general outline unchanged, but with variations in facial aspects meanwhile rising to higher altitudes. At 7·15 p.m. a fine cluster of festoon clouds developed on a prominent outstanding cloud to the west. As the evening became cooler a few light cirro-cumulus clouds came up from the south at 7·35 p.m., the cumulus of the upper strata first dissolved into cirrus when the influence of the southerly reached them, and then quickly dispersed before the wind; lightning visible occasionally to the north until 8·30 p.m. 9 p.m., cirro-cumulus and cumulus extent 4; 10 p.m., overcast nimbus, raining lightly and intermittently until 3 a.m. on 16th, when it came down heavily in Sydney; to south of city it started to rain heavily three quarters of an hour before, and at 2·45 a.m. a heavy clap of thunder was heard with continued rumblings. The rain continued heavy for another hour, when it tailed off.

February 16th—A heavy shower at 9 a.m., light showers during rest of day.

The weather chart of the morning of the 16th February shews that in the previous twenty-four hours the whole storm system had moved about five hundred miles to the east (see weather chart No. 2), the front isobars of the anticyclone overlapping the coast line and the low pressure between Tasmania and the Bluff. The isobars are numerically of the same value as on the 15th, but they have spread out and lost energy, weather generally cooler, with rain in Victoria and on western slopes of the main range in New South Wales.

The second burster to which I shall refer, reached Sydney at 2·15 a.m., February 22nd, 1894, weather chart No. 3 and the accompanying diagram of this period show a great deal more

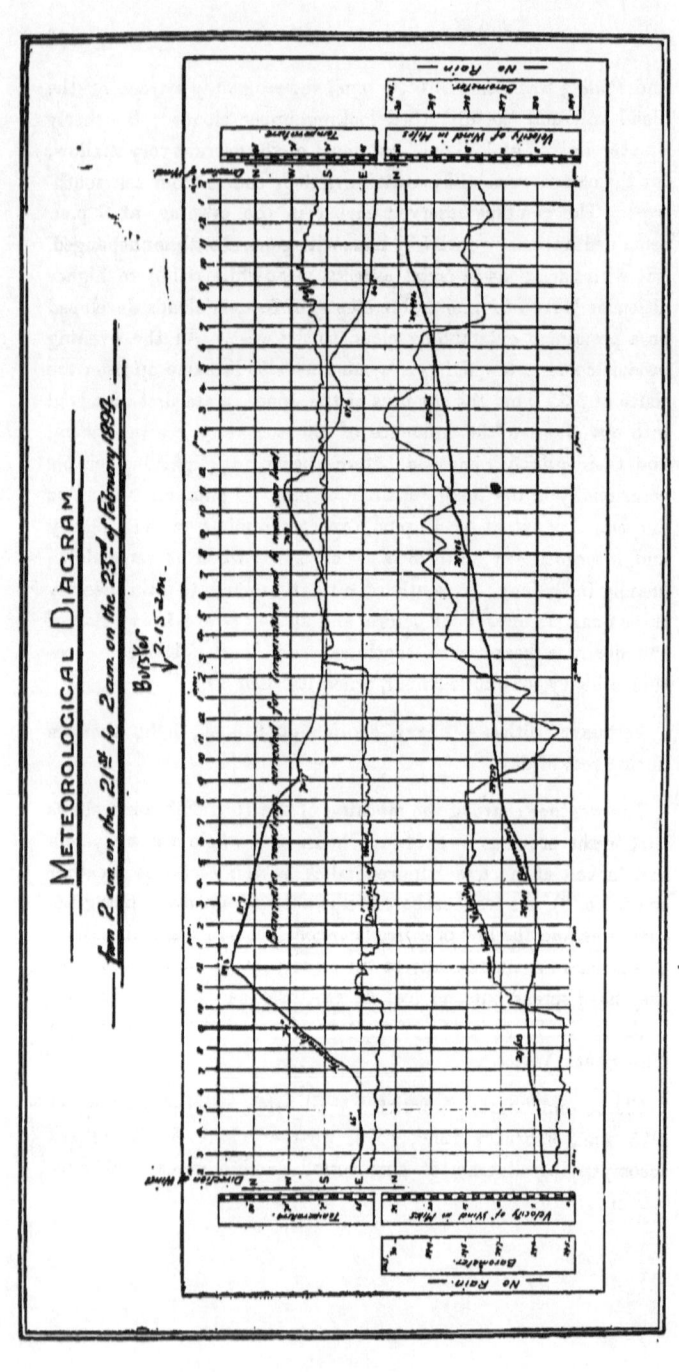

SOUTHERLY BURSTERS.

energy in weather conditions generally than there was on February 15th. The isobars enclosing the approaching anticyclone are more numerous and of higher value. The wind circulation is more regular and moderate to fresh in force generally. The tropical low pressure is also a defined and active feature. The ʌ depression it will be observed on the 22nd was ill-formed, and hence notwithstanding the anticyclonic feature favouring a good blow, this burster was little better than that of the 15th February.

The weather antecedent to this southerly was comparatively speaking cooler in central parts of Australia than on the coast, there was also much thunder cloud, and scattered showers fell in south-west Queensland and northern parts of South Australia on the 21st, these points may in part account for the want of force in this burster. The preceding north-east wind blew for the greater part of three days with a force varying between five and fifteen miles per hour, and at 1 a.m. on the 22nd immediately before the burster it was practically calm. This burster began with a velocity of seventeen miles per hour; between 4 and 11 a.m. the hourly force pulsated between eleven and fifteen miles, it then again rose and twenty-seven miles per hour was reached at 3 a.m., and at 3·15 a.m. thirty-eight miles the greatest velocity; it then gradually fell away and veered through east to north early on the 26th. This may be taken as a modified type of the popular burster. (See page 28).

Cloud notes on burster of 22nd February, 1894—Warm sweltering day on 21st, excepting a small patch of cumulus visible at 3 p.m., not a cloud was seen up to midday 22nd. At daylight on the 23rd cirro-cumulus to extent of ·3 moving from the south, upper strata very light mackerel with just a perceptible motion from west-north-west; clouds gradually extending up to 11 a.m., when it became completely overcast and a light misty rain began to fall at 11·40 a.m. Thunder at 12·50 p.m.; thunderstorm between 1 and 2 p.m., clouds breaking at 3 p.m. to south. An upper strata of cirrus visible to the south-west, but too distant to note motion; cleared by evening.

On November 13, 1893, came a burster of little intensity or interest save from the very remarkable cloud roll, the finest that has been seen in Sydney for some years. The weather conditions at 9 a.m. of November 13 are well shown in weather chart No. 5. An anticyclone rested over Western Australia, with its centre about latitude 30° south, and longitude about seven hundred miles west of Adelaide, while in front of it is a well marked A depression, with its axis north-north-west and south-south-east, and lying centrally over Victoria and the western districts of New South Wales, about three hundred miles west of Sydney. The whole system was moving very rapidly, and from the position of the centre of the high pressure at 9 a.m. of the 14th (weather chart No. 6) the forward motion in twenty-four hours is seen to be fully eight hundred miles. About the A there was a well defined circulation of fresh winds, north-west and south-west, rain had fallen in the northern districts of South Australia and Queensland, and during the forenoon of 13th there were passing thunder squalls and half formed festoon clouds, which were in rapid motion and had a mild disturbed appearance. All day the upper strata of cloud was coming from north-west to west, during the forenoon temperature in the shade rose to 82°. At 3 p.m. seven-tenths of the sky was obscured by cumulus, cirro-cumulus, cirrus, and stratus. At 4 p.m. they were much the same, except that in the north the cumulus clouds were bolder and looked like a wall of rocks, snow-white in colour, with horizontal seams or joints. At 5 p.m. thunder clouds seemed to be in all directions. At 6 p.m. a southerly roll could be seen low down in south-west, with much lightning playing about over it in the cumulo-stratus and cumulus (*Plate*)

As the burster came on, the stratus on the horizon rose gradually up and revealed its true character, at 6·25 p.m. the roll was about 4° in diameter, and had a lower fringe like a very narrow roll of cumulus, above this a dark black roll of stratus surmounted by a feathery fringe, turning up and trending to south, while under the great roll at its northern end, could be seen a well defined shower with rain drifting to north at an angle of 45°. (See *Plate* 4 taken at 7 p.m.) Above this well defined roll cloud another fainter and less defined one is clearly but faintly shown in the negative. The whole storm could be seen to great advantage from the Observatory, and the roll came along getting more and more defined, showing all the features just described from 6·25 to 7 p.m., including the shower and the duplicate roll above. At 7 p.m. it looked very close to us, and could not have been more than three or four miles away, for five minutes later the squall with a velocity of forty-two miles per hour and the rain were upon us. On the photograph taken at 7 p.m., (reproduced as *Plate* 4) the altitude of the lower edge measures 5°, and taking the distance to be four miles, the actual height above the ground comes out one thousand eight hundred and thirty feet, and it spread over at least 100° of the horizon, seventy of which the photograph includes. The sun had set at 6·38 p.m., and night was closing in fast, hastened by the dark masses of cloud which almost covered the sky.

At 6·45 p.m., a long way off and due south, a thunder squall was seen and no doubt marked the arrival of the burster there, showing that the axis of the ʌ was still inclined to south-south-east as it was in the morning. The rate of motion of the axis of the ʌ eastward to Sydney from its barometrically defined position at 9 a.m., three hundred miles west of Sydney, to the Observatory by 7 p.m. *i.e.*, in ten hours, is seven hundred miles in the twenty-four hours, or thirty-three miles per hour.

In the original photograph delicate shading shows the rounded cloud perfectly, with another delicate roll above it, but much of the fine detail is lost in the reproduction.

DANDENONG GALE.

I have also added a diagram, (page 32) giving barometer and anemometer conditions at Sydney during the famous Dandenong gale, for comparison, the extensive and disastrous possibilities which a burster may develop made this desirable. During this gale the wind attained, locally, to the abnormal velocity of one hundred and fifty-three miles per hour in a gust, the rate of one hundred and twelve miles for ten minutes, and fifty-seven miles per hour for nine hours, while its influence was severely felt for hundreds of miles in every direction. The most remarkable points in this diagram are, first the comparatively steady barometer curve; and second, the pulsatory action of the wind velocity shown by the mean hourly number of miles registered during the heaviest part of the gale.

UNFAVOURABLE SEASONS FOR CLOUD OBSERVATIONS.

The opportunities for making cloud observations this summer have been eminently unfavourable, but I have made the best use of such chances as presented themselves, showing the conditions which prevailed for several hours, both before and after the burst. During the present season they have been preceded by clear skies and by overcast skies, and by cirrus and stratus and cumulus, separately and in combination, and to varying extents.

The skies obtaining with the north-westers were more often than not, hazy or overcast, almost tropical of aspect, with hazy stratus and occasional heavy thunder cumulus on the several horizons, but with a special partiality to the south-west.

With the north-easters the skies have been clearer and the smoke-cloud arising from the metropolis, which is seldom noticeable during a north-wester owing to the thick condition of the atmosphere, has generally been perceptible round the horizon in stratified form. The background, or upper strata, when seen, has been composed of fine cirrus, the surface wind occasionally bringing in detached cumulus travelling very low. At times also, as with the north-westers, detached cumulus have been visible to the south-west. The lower strata always moved with the wind, while the upper moved from the west, with an occasional tendency to west-north-west or west-south-west, but this motion could only be discerned in a few instances. In the majority of cases no movement could be detected. Whether the easterly motion is a current or merely the general drift of the atmosphere cannot be decided but as it was noticeable at night I should incline towards the latter hypothesis. No atmospheric current, apart from these two, was observed at any time, except when an agitation was remotely visible overhead between the two, due to vertical uprising or lowering of either one or the other. In these rare instances the clouds moved in all directions.

There were fewer opportunities for observing the upper strata after the burst than while the northerly winds prevailed, but when the conditions permitted observation the cloud movements generally tended from due west. I do not, however, feel inclined to commit myself more definitely on the subject of cloud aspects since the observations from which these statements are deduced are for one season only, and that season an eminently unfavourable one for the collection of useful data.

The question whether bursters arrive here by preference at any particular hour, and whether such hour varies from year to year, as well as the average number for each hour, the percentage of the whole that came at each hour seem to me best answered by a diagram and

TABLE I.

In Table No. 1, all the bursters that have been recorded at Sydney from September 30, 1863 to March 31, 1894, are grouped together so that they can be seen at a glance, all the bursters that

NUMBER IN EACH HOUR AND YEAR.[1]

are on record for each year, and for each hour of the day, also the total number at each hour for the whole period, and the percentage of bursters that have taken place at each hour in figures and in diagram. Each stroke in the table represents a burster, where several bursters have occurred at the same hour there is a group of strokes, and the eye catches at once those hours in each year in which bursters have been most frequent, the greatest

---

[1] See also at the foot of Diagram III., page 58.

number for any hour in any year was between 9 and 10 p.m., 1891, where ten were recorded.

VARY FROM YEAR TO YEAR.

The number for each year varies considerably, the greatest being fifty-six in 1869; the least, sixteen, in 1890. Since 1888 there has been a gradual falling off in the number, and this is coincident with abundance of rain each year.

PREFERENCE FOR A PARTICULAR HOUR.

The Table No. I shows that from 11 a.m. to 1 p.m. bursters are fewest, while they are most frequent from 6 p.m. to midnight, the chances being slightly in favour of 7 to 8 p.m.

It also appears that the hour of maximum varies from year to year, in 1891 it was from 9 to 10; 1888, 11 p.m. to midnight; 1875, 1 to 2 p.m. and so on.

The questions as to the number and strength of bursters in the several seasons and months of the year I have treated in the same manner.

TABLE II.

In Table No. 2, all the bursters on record have been grouped in another way, which brings into evidence other characteristics. The table has the years on the sides and months across the top, and the information given under each month and for each year is the maximum velocity of the wind, and if there are several bursters in the month the average of their greatest velocities; at the right hand side the average of the greatest velocities for the year, and the greatest velocity attained at any time during the year. It is also shown that the average number in each year is thirty-two, the greatest in any year fifty-six, and the least sixteen. The average of all the greatest velocities is 42·7 miles per hour, the greatest one hundred and fifty-three miles per hour. (It must be borne in mind that all the anemometer results at Sydney are recorded on the assumption that the velocity of the wind is three times that of the centres of the cup, this has been the practice since 1862, and it is deemed better not to alter it until the exact ratio has been decided and generally accepted.)

It is necessary to state here that in the conditions under which this prize essay was written, there is no express definition of the minimum velocity of wind which shall constitute a burster, I have therefore taken as such every sudden shift of wind to south or south-west, from any point between west and north on the one side and east and north on the other side, irrespective of the velocity of the wind; always provided however, that the velocity was maintained subsequently for some hours with force.

It will be observed that the table brings to light the fact, that the mean velocity of the wind is greater in spring than in autumn, which may be accounted for by the fact that our September equinox is windy and the March one wet. The discrepancy between the seasonal total in the tables is accounted for by the fact that the anemometer on a few occasions failed to record the hours of arrival, in these cases the burster is omitted from Table I.

One other point seems important in respect to the arrival of bursters, and that is the time between the lowest barometer and the arrival of the burst.

DIAGRAM III.

Gives the fluctuations of the self recording barograph for twelve hours before and after nineteen bursters of the summer 1893 to 1894. The curves have been drawn with their lowest points in a vertical line, and the times at which the burst commenced on each curve have been connected by lines, thus showing the relation between the barometic minimum and the time that the burst of wind reached Sydney. It will at once be seen that the bursters occur some hours after the barometers have commenced to rise, and if this set of curves be accepted as the general rule, the diagram shows that in any season the lowest readings take place more remotely from the change of wind, early in the summer. It is also patent that no particular height of barometer is peculiar to the burster. A sudden rise does not indicate that the blow will be a hard one, see the sharpest that of 17th September and of 30th November, while the velocity of forty-eight miles per hour,

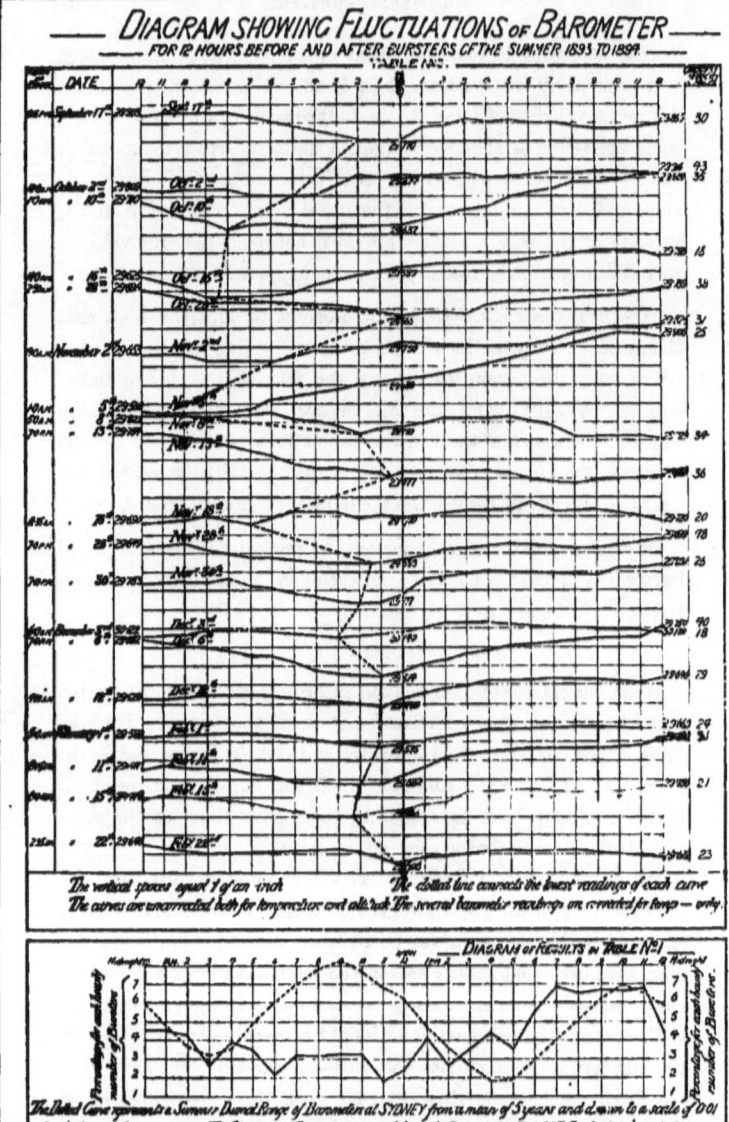

SOUTHERLY BURSTERS. 59

the heaviest of the season resulted from a very gradual rise and with a range of only ·1 in the whole twenty-four hours. The curves at the foot of diagram 3 relate to one feature of Table I.

HOT SOUTHERLY WINDS.

Instances have been recorded of hot southerly winds which lasted for some hours. These are no doubt, the winds blowing on the western aspect of the tropical depression. The following is an account of a hot southerly wind copied from notes made by Mr. Russell, and kindly placed at my disposal :—

"December 12th, 1883—Yesterday was hot, to-day hotter still, and by 9 a.m. the temperature had risen to 82·6, wind from north and hot. At noon the humidity fell to twenty-nine, and the temperature had risen to 99·6. At 3 p.m. the dry bulb stood at 95·6 and the wet at 71·1 ; from 9 a.m. to 6·30 p.m. the barometer fell 0·350 and then began to rise rapidly, and by 8 p.m. had risen 0·215. At 6·30 p.m. as the barometer began to rise the wind changed suddenly to south, and it felt more like a furnace blast than a southerly burster. The wind continued as a hot wind for more than an hour, but not strongly after the first violent gusts were over ; by 9 p.m. a light warm air from south-west had set in but the temperature was still 86·6. I do not remember a hot southerly like this before. On the morning following we had a cool wind from north-west."

In conclusion, it is needless to state that this investigation has involved much labour and a vast amount of patient research, and though I will be gratified if my efforts should in some degree, however small, contribute to a better knowledge of this phenomenon : in any case, I will consider myself rewarded for the labour I have given to is by the knowledge I have acquired while gathering material for this essay. I am largely indebted to Mr. H. C. Russell, who has kindly given me access to all the valuable records of the Sydney Observatory, and furnised me with copies of weather maps, photographs, etc., for which my warmest thanks are hereby expressed.

AUSTRALIAN WEATHER.

TABLE II.—TABLE SHOWING NUMBER OF BURSTERS IN EACH

Year.	August.			September.			October.			November.			December.		
	Number	Velocity.		Number	Velocity.		Number	Velocity.		Number	Velocity.		Number	Velocity.	
		Mean	Extreme		Mean	Extreme		Mean	Extreme		Mean	Extreme		Mean	Extreme
Summer.					Rate in Miles per hour			Rate in Miles per hour			Rate in Miles per hour			Rate in Miles per hour	
1863-64	...	...	...	2	58·0	70	6	54·8	108	7	29·4	41	3	32·3	40
1865	...	...	...	2	11·0	12	3	32·3	49	5	25·6	50	4	24·3	36
1866	...	...	...	2	26·5	32	5	43·8	64	10	38·9	74	7	28·3	48
1867	1	20·0	20	6	25·7	40	...	...	...	6	28·7	42	4	38·0	56
1868	...	...	...	3	30·0	47	3	23·3	29	10	29·4	43	4	36·3	69
1869	...	...	...	5	17·4	27	7	43·0	60	11	37·0	72	8	37·8	73
1870	1	18·0	18	2	29·0	36	4	32·5	47	4	44·3	70	8	30·5	53
1871	...	...	...	1	39·0	39	7	33·8	52	6	40·0	69	6	45·0	63
1872	...	...	...	1	37·0	37	5	35·6	54	5	30·2	40	6	42·5	64
1873	...	...	...	2	40·0	42	2	40·0	44	1	47·0	47	4	29·3	45
1874	...	...	...	...	...	...	5	36·8	54	4	34·7	44	6	37·3	61
1875	...	...	...	1	20·0	20	6	40·8	54	5	29·4	41	8	32·9	54
1876	2	33·5	37	3	24·3	30	6	28·3	34	6	26·7	37	4	31·0	50
1877	...	...	...	6	47·2	153	5	34·8	43	6	30·8	55	7	32·9	54
1878	...	...	...	2	29·0	32	3	46·3	59	8	40·0	53	9	40·1	64
1879	...	...	...	4	29·3	42	7	45·9	72	9	30·5	45	9	31·0	52
1880	...	...	...	...	...	...	4	39·2	76	5	35·4	63	8	37·8	61
1881	...	...	...	2	28·5	33	7	38·3	59	6	40·0	68	8	61·1	82
1882	...	...	...	2	36·5	41	3	47·0	63	5	42·0	54	5	29·6	48
1883	1	16·0	16	4	43·8	68	5	34·2	59	4	34·3	53	7	32·1	43
1884	...	...	...	1	32·0	32	5	34·4	45	4	22·5	27	7	30·6	54
1885	...	...	...	5	28·8	48	7	35·4	41	5	34·0	48	5	38·2	50
1886	...	...	...	1	30·0	30	8	26·9	36	7	33·0	47	4	38·0	41
1887	...	...	...	...	...	...	4	32·5	63	3	24·3	29	4	30·3	37
1888	...	...	...	1	43·0	43	3	35·3	37	4	34·5	48	5	28·0	50
1889	...	...	...	1	18·0	18	4	34·0	43	3	35·0	46	6	29·0	43
1890	...	...	...	1	21·0	21	1	26·0	26	3	19·7	24	8	29·1	42
1891	...	...	...	1	21·0	21	2	27·0	35	2	31·0	48	4	22·5	30
1892	...	...	...	...	...	...	2	21·0	21	3	35·7	41	5	28·8	35
1893	...	...	...	...	...	...	6	20·7	48	2	33·0	42	6	32·3	58
1894	...	...	...	1	30·0	30	4	33·5	43	7	31·4	48	3	29·0	40
Total	5	...	...	62	...	...	139	...	...	166	...	...	182	...	...
Percent.	·5	...	...	6·3	...	...	14·0	...	...	16·8	...	...	18·3	...	...
Mean	·2	21·4	...	2·0	30·3	...	4·5	35·5	...	5·4	33·5	...	5·9	33·7	...
Extreme	2	33·5	37	6	58·0	153	8	54·8	108	11	47·0	74	9	45·0	73
Year	...	1876	1876	...	1864	1876	...	1864	1864	...	1873	1866	...	1871	1860

Note.—In making out this Table the greatest velocity of each southerly has if there are several bursters in the month, the average of their greatest veloci-

MONTH WITH MEANS AND EXTREMES OF GREATEST VELOCITIES.

January.			February.			March.			April.			May.			Total Number	Velocity.
	Velocity.			Velocity.			Velocity.			Velocity.			Velocity.			
Number	Mean	Extreme	Number	Mean	Extreme	Number	Mean	Extreme	Number	Mean	Extreme	Number	Mean	Extreme		Mean
Rate in Miles per hour			Rate in Miles per hour			Rate in Miles per hour			Rate in Miles per hour			Rate in Miles per hour				Rate in Miles per hour
7	22·0	48	4	30·5	31	2	34·0	34	...	...	...	...	...	...	31	37·3
8	29·5	45	7	22·4	36	4	24·5	38	3	27·3	40	...	...	...	36	24·6
6	29·8	46	3	40·5	46	2	50·0	54	2	22·5	23	...	...	...	37	35·0
5	48·4	69	6	26·7	44	4	16·0	18	5	26·0	26	...	...	...	37	25·3
8	35·4	54	3	51·0	77	4	33·3	63	3	14·7	16	3	14·5	15	38	31·7
7	34·4	63	5	24·0	33	8	34·3	61	2	24·0	36	...	...	...	56	29·6
5	39·8	49	5	41·6	55	...	...	...	...	...	...	1	43·0	43	29	33·6
3	35·3	52	4	38·8	51	3	19·0	25	3	40·0	56	...	...	..	34	37·1
6	34·0	50	7	33·4	40	1	40·0	40	3	19·3	24	...	...	...	34	34·0
6	35·7	45	2	29·5	34	5	20·4	27	...	...	...	...	...	...	22	34·4
6	34·3	47	5	28·8	44	1	35·0	35	3	20·0	29	...	...	...	30	32·4
9	38·1	52	3	38·0	64	5	29·0	36	1	23·0	23	...	...	...	38	31·4
4	32·3	44	5	32·4	39	3	35·0	43	2	29·0	35	...	...	...	35	30·3
7	26·0	38	4	26·0	34	1	39·0	39	1	44·0	44	...	...	...	37	35·1
4	35·3	44	2	39·0	41	4	19·7	24	...	...	...	...	...	...	32	35·6
7	39·0	52	4	31·3	36	3	34·7	47	4	27·5	46	...	...	...	47	33·7
7	29·0	36	1	36·0	36	3	30·0	32	...	...	...	...	...	...	28	34·6
5	48·0	71	5	34·6	54	3	27·3	30	1	64·0	64	...	...	...	37	42·7
8	37·6	67	5	30·8	66	4	34·7	45	2	20·0	20	...	.·.	...	34	34·8
5	34·0	50	3	28·7	37	1	28·0	28	...	...	...	...	...	...	30	31·4
8	39·3	54	7	29·9	46	3	33·7	43	...	...	...	...	...	...	35	31·9
6	29·3	40	7	30·1	50	·7	28·7	41	1	20·0	20	...	...	...	43	28·5
4	25·5	30	5	35·8	46	3	39·0	47	...	...	...	...	...	...	32	32·6
2	30·5	32	1	41·0	41	4	27·8	34	2	34·0	42	...	...	...	20	31·5
8	29·8	46	5	31·6	45	4	27·0	30	2	23·5	31	...	...	...	32	31·6
5	27·8	35	5	31·0	34	3	24·0	33	...	...	...	...	...	...	27	28·4
3	37·3	40	...	...	...	...	...	...	...	...	...	...	...	...	16	26·6
6	26·3	40	7	24·1	42	1	27·0	27	...	...	...	...	...	...	23	25·6
2	40·5	46	5	22·0	26	3	25·7	27	1	21·0	21	...	...	...	21	27·8
3	33·0	57	3	24·0	35	1	30·0	30	...	...	...	...	...	...	21	30·3
...	...	...	4	27·0	30	...	...	...	...	...	...	...	...	...	19	30·2
170	...	...	132	...	...	90	...	...	41	...	...	4	...	...	991	...
17·3	...	...	13·2	...	...	9·1	...	...	4·1	...	...	·4	...	...	100	...
5·5	33·9	...	4·3	31·7	...	3·2	30·2	...	1·3	27·7	...	0·1	28·8	...	32	31·9
9	48·4	71	7	51·0	77	8	50·0	63	5	64·0	56	3	43	43	56	42·7
...	1867	1881	...	1868	1868	...	1866	1868	...	1882	1871	...	1871	1871	1869	1881

been found by ascertaining the shortest time in which four miles of wind was re
ties is shown under the head "Mean."

Years	A.M. 0–1	1–2	2–3	3–4	4–5	5–6	6–7	7–8	8–9	9–10	10–11	11–12	P.M. 12–1	1–2	2–3	3–4	4–5	5–6	6–7	7–8	8–9	9–10	10–11	11–12	Total Year
1863–64																									31
1865																									36
1866																									37
1867																									32
1868																									38
1869																									56
1870																									29
1871																									34
1872																									33
1873																									22
1874																									30
1875																									38
1876																									35
1877																									37
1878																									33
1879																									47
1880																									28
1881																									37
1882																									33
1883																									28
1884																									35
1885																									42
1886																									32
1887																									20
1888																									32
1889																									27
1890																									16
1891																									23
1892																									21
1893																									21
1894																									19
Total	44	45	41	28	30	34	21	30	32	31	17	21	41	27	34	43	34	54	69	65	67	66	68		981
Percentage	4·5	4·6	4·2	2·8	4·0	3·5	2·1	3·1	3·3	3·2	1·7	2·1	4·2	2·7	3·5	4·4	3·5	5·5	7·0	6·6	6·8	6·7	6·9		Mean Yrs. 31·6

TYPES OF AUSTRALIAN WEATHER.

By HENRY A. HUNT,

Second Meteorological Assistant, Sydney Observatory.

[With Forty Diagrams.]

In continuation of the valuable work on Australian Meteorology which the Hon. Ralph Abercromby initiated several years since, by offering a money prize for the best essay on Southerly Bursters; he has recently selected the phases of Australian Weather which are treated in the following twenty studies of "Types of Australian Weather." Many of these appear to be peculiar to Australia, and at the same time connected with Equatorial and other weather. That they throw much new light upon the source of the greater part of Australian rain, and show how these rain storms develop out of ordinary weather conditions is certain; at the same time they form an important contribution to the study of weather in the Southern Hemisphere generally. The work has been done by Mr. H. A. Hunt, at Mr. Abercromby's expense, and Mr. H. C. Russell, has edited the work.

GENERAL REMARKS.

As a general rule, weather is set fine when anticyclones move rapidly, and in a straight line across Australia, *i.e.*, at a rate exceeding five hundred miles per day. And weather is unsettled when they move slowly, and not in a straight line, *i.e.*, in a zigzag line, especially if they show no appreciable forward motion for a day or two. When anticyclones move in low latitudes the conditions favour dry weather, in high latitudes, wet weather, especially if they rest for a time south of South Australia.

All the examples of weather phases which follow have been selected from the Sydney Weather Charts, and illustrate each type. The originals were carefully traced, and then reduced from $22'' \times 17''$ to the size used in this essay by means of the camera.

Ordinary symbols have been used, except the circle half filled for thunderstorms, and the straight line shading; parallel lines indicate the area of rainfall under one inch, crossed shading over one inch.

TYPE I.—MOVING ANTICYCLONES.

One of the best marked features of Australian weather is the steady easterly progression of all the types, and the governing type, that in fact about which all the other types seem to congregate, is the anticyclone; it has therefore been placed first in the series, with three charts to show the progress made by a quick moving one in forty-eight hours. The average daily progress of anticyclones is four hundred miles per day, but the speed at times rises to one thousand miles.[1]

Investigation so far leaves no room to doubt that in these latitudes a series of anticyclones surround the globe; the latitude of the average one varies with the season, being farther south in summer than in winter. The normal circulation about an anticyclone brings southerly winds in front of them, and northerly winds in the rear, hence our cold and our hot winds.

Chart No. 1 shows the position, on 15th August, 1893, of the eastern half of an incoming anticyclone; it rests over Western Australia, while the departing one is seen over the Tasman Sea; between these is seen the usual ∧ depression, which is of average intensity, and a dormant tropical low pressure to the north. In Chart No. 2 the anticyclone has moved nearly nine hundred miles in the twenty-four hours, the centre being located near Fowler's Bay, north of the Australian Bight; the antarctic ∧ depression is well across the Tasman Sea, while the tropical or monsoonal isobar depicted in the previous chart has apparently merged into the high pressure system, a curious and not unusual kink being formed to the north-east of it, following the contour of the Gulf of Carpentaria. On Chart No. 3 the anticyclone is shown to have moved a further seven hundred and fifty miles, or a total of

---

[1] Russell—Quarterly Journal R. M. S., Vol. XIX., No. 85.

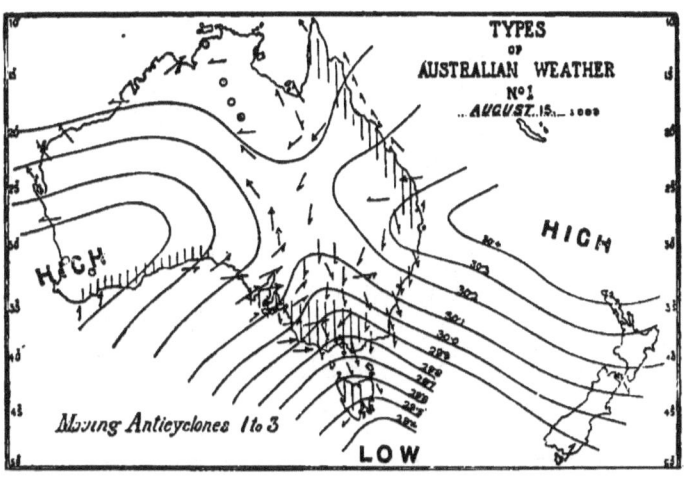

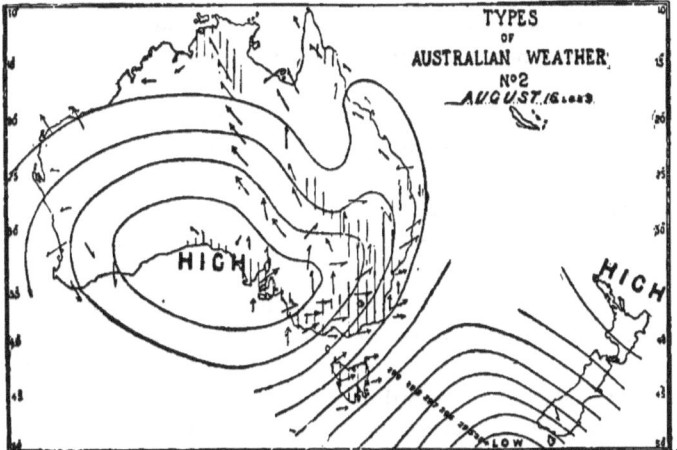

one thousand six hundred and fifty miles in forty-eight hours; the monsoonal dip is still noticeable south of the Gulf, while the antarctic Λ depression has its centre to the east of New Zealand.

E

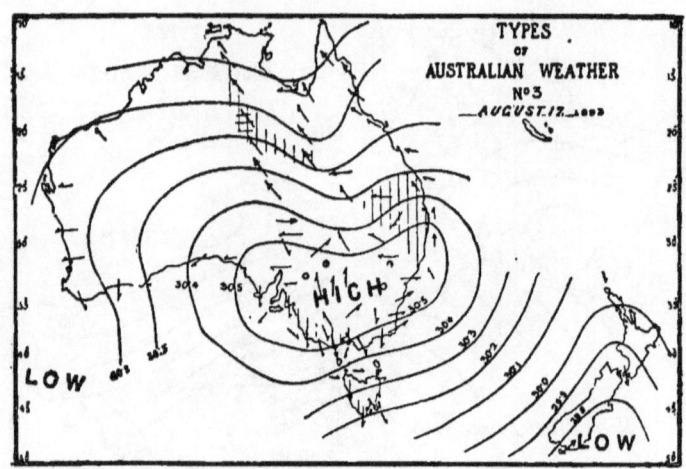

The passage of this anticyclone was an unusually rapid one, and is presented as portraying with a minimum number of charts the easterly motion of anticyclones over Australia.

TYPE II.—MONSOONAL RAIN STORM.

This type is undoubtedly the chief rain agent in the Australian Continent. Monsoonal depressions or tongues may occur at any time of the year, but particularly between the months of September and April, and most frequently during January, February and March. The readings of barometers in the depression seldom fall very low, the grade from the surrounding areas to the centre of the tongue ranging from one to three-tenths of an inch generally; the depression may intensify, that is the tongues between high pressures may protrude further south anywhere during their passage across Australia, but show a preference to do so after they have crossed central Australia, a fact which suggests that the heated interior has at least some influence in their development.

When and wherever the tongue is well defined, rain certainly follows in its track, and thunderstorms as a wide spread and simultaneous feature are never experienced without it.

On April 18th and 19th, 1894 (Chart Nos. 4 and 5) occurred one of the finest monsoonal rain storms on record, the area affected being very extensive, embracing the whole of the eastern colonies, Tasmania, and the greater part of South Australia. Many inches —up to five and six—of rain were recorded on the north-east coast of Queensland, and over the eastern half of the rain area the country benefitted to the extent of over two inches generally before the storm was over.

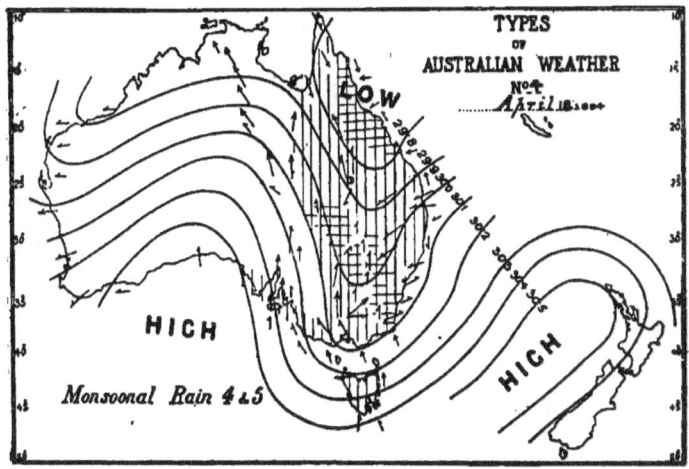

On this Chart No. 4 the monsoonal isobars, after passing the most southern part of the tongue, are shown sweeping round a high pressure of considerable energy, situated over the Great Bight. Following these isobars to the eastward, we find them recurving over another high pressure of greater energy in the Tasman Sea. On Chart No. 5, both anticyclones will be noticed to have worked northward, and while so doing they have lost somewhat in pressure, and the low pressure tongue has extended further south and broadened; at the tip, two cyclonic rain centres will be observed to have formed.

The chart antecedent to 18th April presented the general characteristics of Chart No. 1, except that the antarctic ∧ depression was a considerably *less active* feature, but on the other hand the monsoonal dip was somewhat more pronounced, and possessed two instead of one isobar. Following the 19th, the chart of 20th revealed no sign of the monsoonal tongue, and the continent was covered with a high pressure of very slight energy.

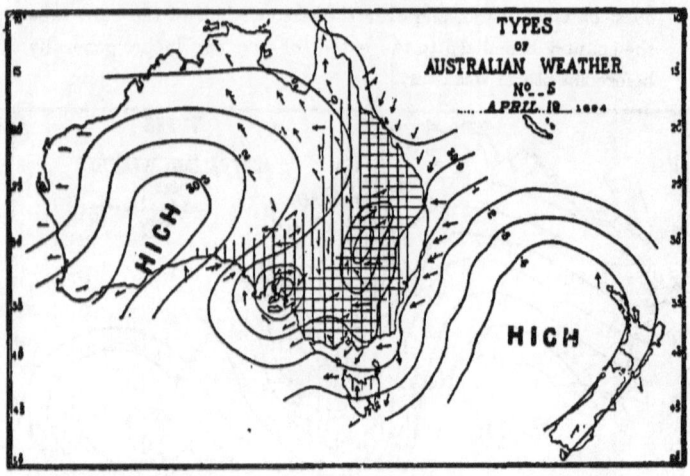

TYPE III.—DEVELOPMENT OF A CYCLONIC STORM IN LOW LATITUDES FROM A MONSOONAL DEPRESSION.

In Type No. 3 we have the development of a cyclonic storm out of a monsoonal depression. The seasonal peculiarity of this phase of the tropical low pressure is similar to that in Type No. 2. The cyclone seems to develop when the southern extension of the monsoon is out of proportion to its width, and it becomes so narrow at one part of it that the opposing winds which circulate round it interfere, set up the cyclonic circulation, and it then progresses eastward as a rain storm. (See Charts 6 and 7.)

These storms frequently develop in South-east Queensland, and they are generally most severe there; the quantity of rain which sometimes comes with them is very remarkable, as in the case of the phenomenal flood in Brisbane in 1893, which was the result of one of these storms.

The example selected and shown in Charts 6 and 7 took place on September 28th and 29th, 1892; on the 28th (Chart 6), it appears that one of these storms had developed in the previous twenty-four hours, at the southern end of a narrow tropical or monsoonal tongue of low pressure. An anticyclone of good energy lies to the east of it, with its isobars extending well to the north and contracting the width of the tongue in the north; another anticyclone lies to the west, and this also seems to be pressing on the narrow monsoonal tongue and helping to contract its diameter, at the same time the energy in both high pressures is adding force to the circulation of the wind, and so aiding in the development of the cyclonic circulation. A modified $_\Lambda$ depression exists to the south-east of the Australian Bight, and there is another over New Zealand.

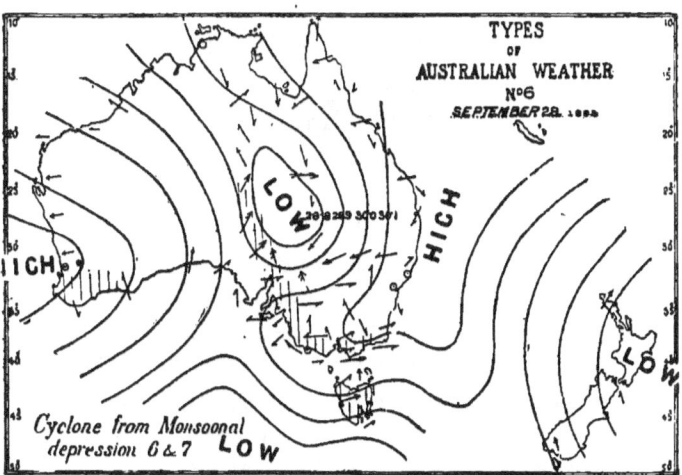

On September 29th, (Chart 7), the small cyclone has extended its area and energy on its way to the coast, but its motion of rotation has been rapid, and this probably accounts in a measure for the comparatively small area over which rain has fallen, although, in this instance, over an inch of rain fell in the central and eastern parts of New South Wales. The winds circulating about the western isobars of this storm are rather stronger than usual in such cases. In the twenty-four hours both anticyclones have lost a considerable portion of their energy, while that of the depression in New Zealand is about the same as on the previous day.

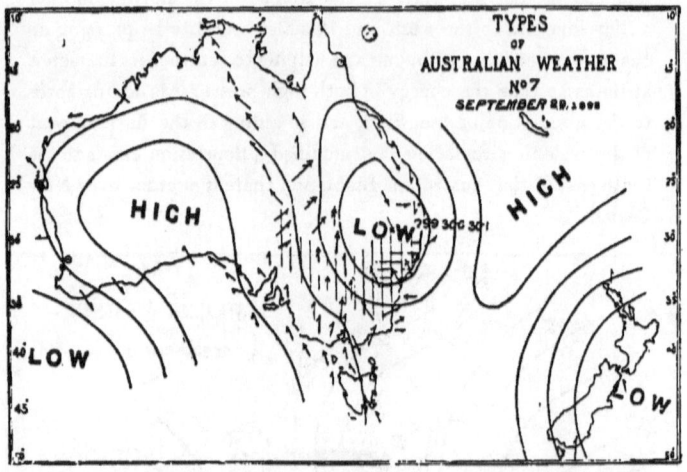

TYPE IV.—DEVELOPMENT OF A CYCLONIC STORM IN HIGH LATITUDES FROM A MONSOONAL DEPRESSION.

These are somewhat similar to Type 3, but the rainfall is usually not so heavy, and the wind much more violent.

Chart No. 8 shows the development of one of these cyclones on April 15th, 1889; in this case the monsoonal depression had extended across Australia into the Australian Bight. The development of one of these storms is heralded by the strong easterly gales on the south-east coast of South Australia and south coast of Victoria. The energy of wind circulation increases over South Australia, and as the whole system moves bodily eastward from the Australian Bight to the mainland, the circulation is seen to be that of a fully developed cyclone of small area, with a diameter three to five hundred miles, formed out of the southern part of the extensive monsoonal depression.

In Chart 8 a high pressure is shown over the south-western part of Australia, and evidently encroaching on the moonsoonal about central Australia.

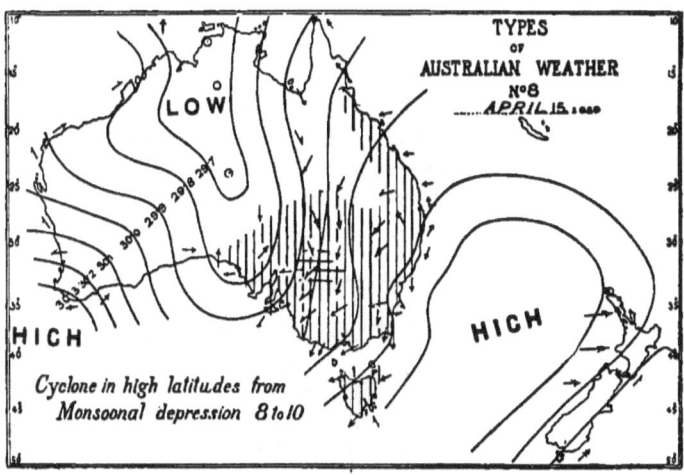

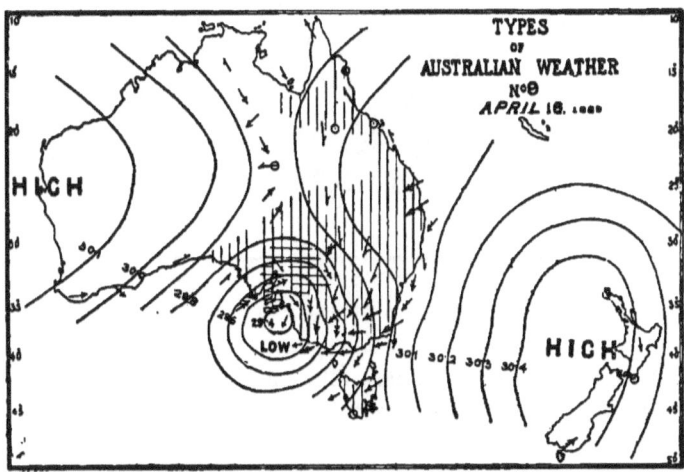

In Chart 9, the western anticyclone has extended northwards and has seriously contracted the diameter of the monsoonal low pressure, thereby facilitating or helping to cut off and start the cyclonic storm in the south; which has now developed a much steeper barometric grade and an energetic wind circulation, and is moving eastward without extending the rain area.

Chart No. 10 shows comparatively little motion in the cyclone, but it is considerably distorted, especially to the west, where the anticyclone is compressing the isobars by its easterly progress and in its endeavours to maintain its rate has overridden the cyclone in the north. As will be seen by a glance at these three charts, the rain resulting from this storm was most extensive and beneficial, and the winds under its influences were strong.

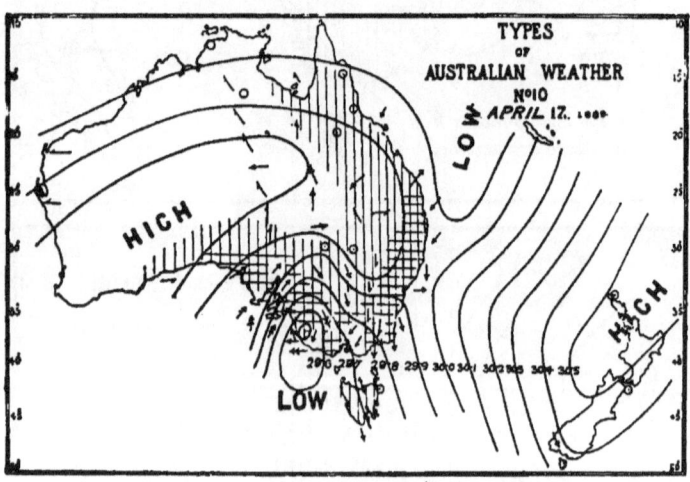

TYPE V.—CONDITIONS FAVORABLE FOR THUNDERSTORMS.

Upon comparison of charts setting forth this type with those of cyclonic thunderstorms, they will be found very similar; the main difference being the absence in this set of the cyclonic area at the end of the monsoonal tongue. The chief feature indicating thunderstorms is the narrowness of the col[1] and somewhat congested state of isobars in the high pressures west and east of it, resulting in opposing winds. Those of the tropical tongues are hot and charged with moisture, while those of the Λ depressions are strong and dry. This type is met with during the monsoonal season. The rains resulting from this feature are not generally heavy, and though thunderstorms may be experienced over extensive areas of Queensland, New South Wales, and Northern Territory, a number of them may occur without any rain falling.

Chart No. 11, January 17th, 1893, shows an extended and narrow tongue lying between two relatively high pressures, a Λ depression similarly situated also exists in the south, this col or area of low pressure separating these two depressions is very small, and opposing currents of wind are noted there as blowing within close limits. Thunderstorms were occurring or had occurred in Northern Territory generally, over a great part of Queensland and in central parts of New South Wales, the rains upon this occasion, as is often the case, were not heavy, though they fell over an extensive and generally unfavoured area.

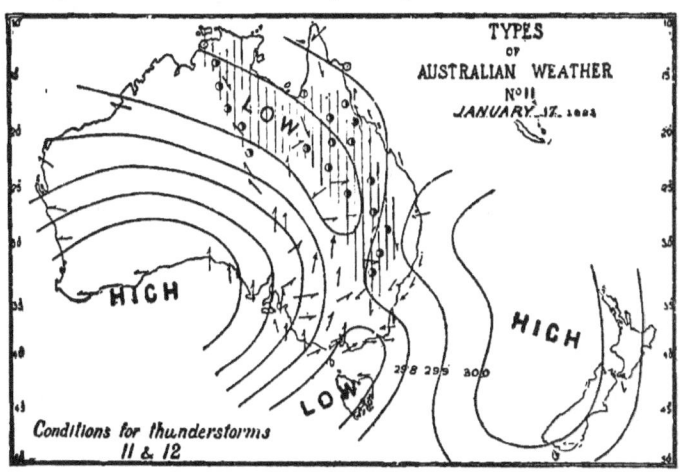

---

[1] For explanation of this term see p. 80.

On Chart No. 12, the following day, though a few storms were recorded, shows a great diminution in number. The col has widened and the monsoonal tongue has lost to some extent its thundery characteristics, having widened at the end. The accompanying barometric systems show no motion since the previous day, but the high pressures have intensified.

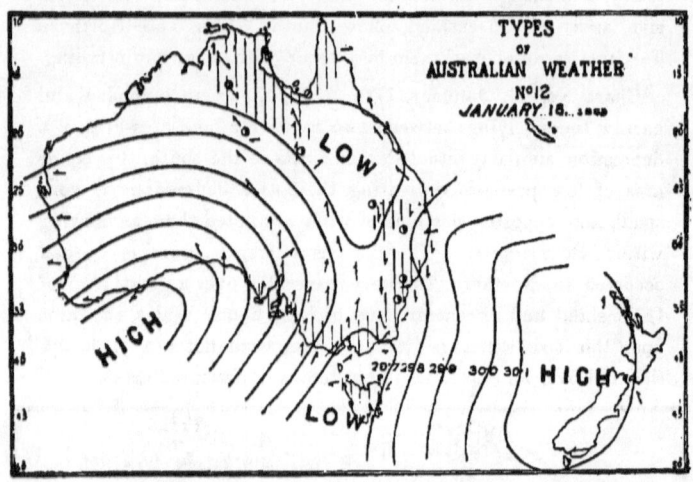

TYPE VI.—CYCLONIC THUNDERSTORMS.

This also, like the preceding one, is allied to the tropical low pressures, but in this case a defined cyclonic circulation develops in the lower extension of the tongue without the usual intensification of grades. From this source the thunderstorms radiate in easterly and southerly directions, and at times, as in the instance presented, a vast area is affected.

Chart No. 13, December 12th, 1893. As on the previous set the monsoonal tongue lies over much the same country, though with its axis more east and west; the high pressure to the west is very small, and the systems are somewhat fragmentary.

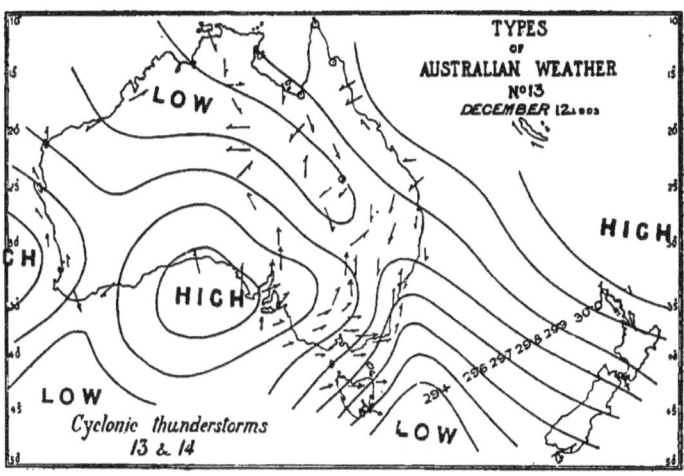

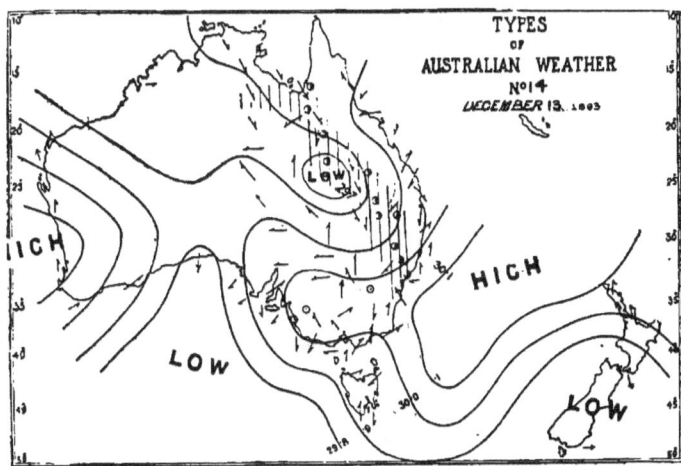

On the following day, Chart No. 14, a cyclonic circulation has formed in the end of the tongue, while the Λ depression over Sydney on the 12th has moved eastward and nearly filled up, and the high pressures have followed and intensified. The thunderstorms and rains were not so general as on the thunderstorm type, but curiously, were recorded in an almost parallel area of country, nearly three hundred miles wide by twelve hundred long. Though great barometric changes have taken place, the pressures remain about the same as on Chart No. 13.

TYPE VII.—A RAIN STORM WITH VERTICAL AND NEARLY STRAIGHT ISOBARS.

This type is one of the best defined and reliable of the series for forecasting purposes, because with them, good general rains almost invariably come. They are the rear isobars of a departing anticyclone, and the wind circulation from north and north-east brings into the interior winds laden with tropical moisture to meet in the west southerly winds laden with antarctic cold, and therefore precipitating power. The sequence of rain is rendered even more certain if rain be recorded in the north-east before the isobars straighten, or in the $\wedge$ depression to the south.

The actual height of the barometers is not material, but the greater the number of isobars in a given area, the more extensive will be the rainfall; the rain usually lasts three days. The rain begins to fall north-west of New South Wales, spreads southwards, then eastward, and finally northwards, crossing the mountains near the Queensland boundary. A fine example of this type occurred on 15th, 16th, and 17th October, 1894. (Charts 16, 17.)

On the 15th, Chart 16, a departing anticyclone rests over Tasman Sea, and its rear isobars are shown running north and south over central Australia; another anticyclone is shown over Western Australia and a $\wedge$ depression east of the Australian Bight; a trough of low pressure rests over Central Australia from north to south. On this day the only indications of the 'pending rain were found in the cloudy skies generally over South Australia, western parts of New South Wales and Queensland, and a small area of rain in South Australia.

In Chart 16, the straight isobars of Chart 15 have entirely disappeared, but the rain has come over Central and South

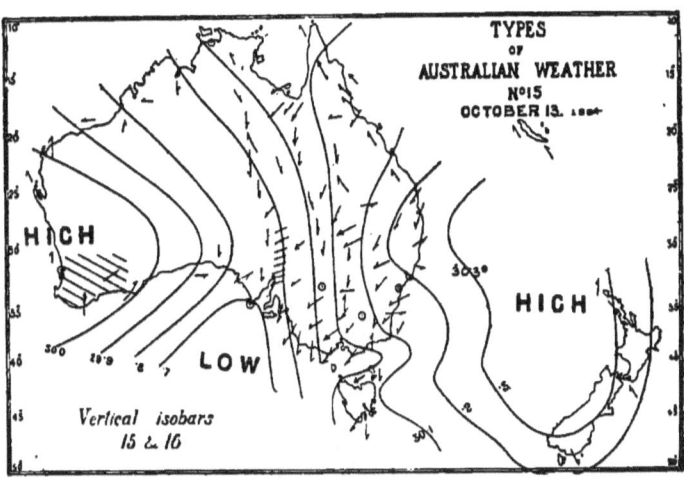

Australia. Excepting the Gulf country and its central area, all Queensland, all New South Wales, Victoria and Tasmania had rain. (See shading on Chart 16). For other instances of straight isobars see Charts 5, 27, and 28.

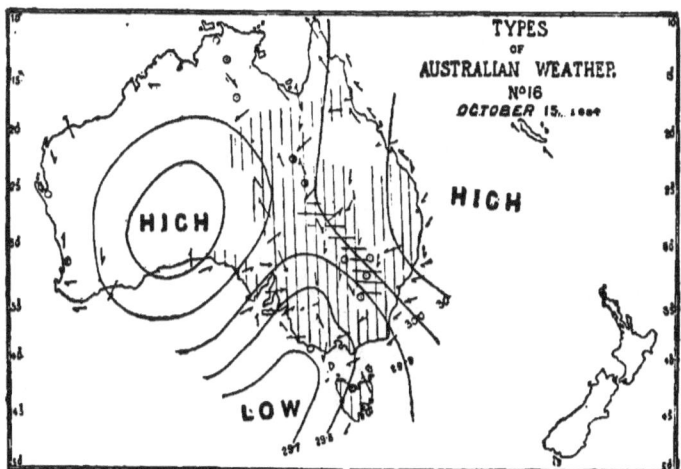

TYPE VIII.—CYCLONES FROM NORTH-WEST.

From time to time fully developed cyclonic storms appear on the north-west and west coasts of Australia, and in the Australian Bight, but the absence of observing stations in the unoccupied country which lies between the overland telegraph line and the west coast of Australia, makes it impossible to trace them over that part of the continent, but cyclones are well known on the northern coast of Western Australia, and their formation in the tropics equally well known. There can therefore be no doubt that when we find a cyclone on the western coast of Australia or in the Australian Bight, that it is one which has come from the north-west, and is in fact recurving to the east and south-east as they do on the east coast.

The one selected for Type 8 was picked up on the west coast of Australia in Latitude 28° on July 4th, 1892, (See Chart No. 17). The winds were light, but the rain heavy along the coast; an inert anticyclone rested over South Australia, Victoria, and New South Wales, where it was moving to the east and making room for the approaching storm.

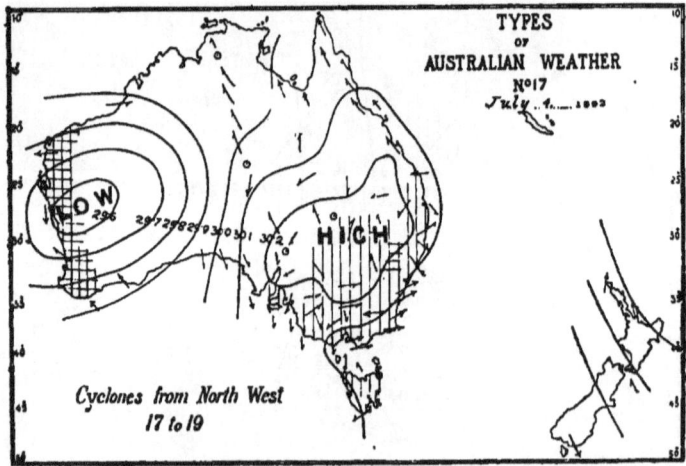

Chart 18, July 5th, shows that the whole system has moved rapidly, that the cyclone now with an elongated centre lying north-west to south-east now rests over South Australia, strong winds are rapidly developing, and the barometric grade about the

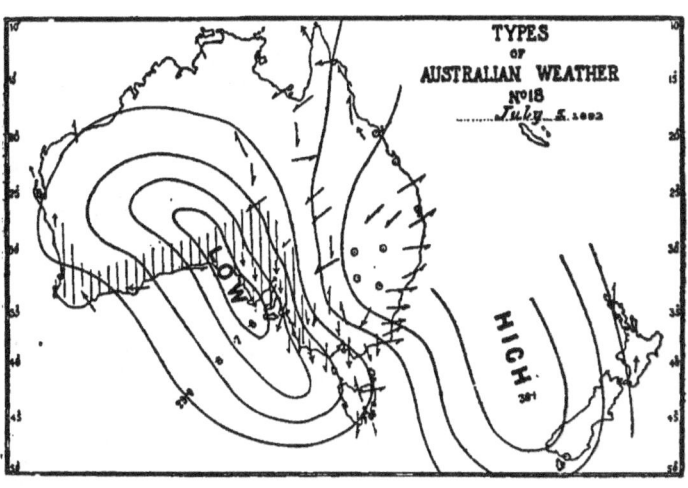

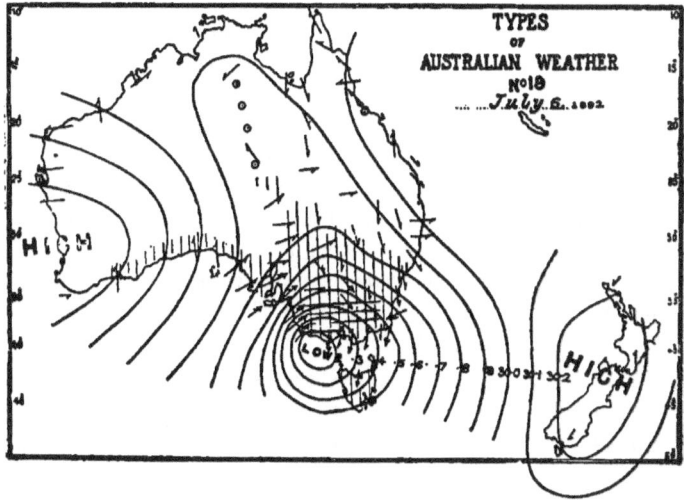

centre is much steeper, and rain has fallen over the south coast
generally and extended northwards almost to Central Australia.

On July 6th at 9 a.m. the weather chart presented the features
shown in Chart 19. The cyclone has intensified all round, and
has moved rapidly to the east, its centre is just entering
Bass' Straits; very heavy gales from south-west are blowing
in the rear of the centre; heavy rain is falling over Victoria and
extends over the greater part of New South Wales. All the winds
controlled by this storm were very heavy, and during the 6th
July, as the storm passed through Bass' Straits, extremely heavy
weather was experienced there. On July 7th it had filled up.

TYPE IX.—TORNADOES.

These occur during the summer months, and are most frequent
in the western plains; they are developed in hot weather and in
the low pressure known as a "Col" between two high pressures,
when there is not enough grade to control the winds and the heating power of the sun is great; if to these there is added the pre-
sence of moisture from recent showers, we have all the conditions
for the formation of a tornado. The force of wind is often sufficient to break off growing trees, two and even three feet in
diameter, and the reason there is so little damage to life and
property is not the want of power, but the sparse population and
the very small number of towns.

Chart 20, March 20th, 1894. An extensive anticyclone lies
south of Australia, giving way in its central parts to an extensive
monsoonal dip. The isobars are generally uniform and of even
gradients, though a suspicious interval exists to the west of New
South Wales between the 30·0 and 29·9 curves. This is undoubtedly the area in which the secondary developed. Light rains
were recorded in New South Wales and Victoria, but in Central
Australia temperatures were high.

Chart 21, March 21st. A marvellous change has taken place.
The area of high pressure on the previous day over the Tasman
Sea has lost two-tenths in pressure. The monsoonal dip so pro-

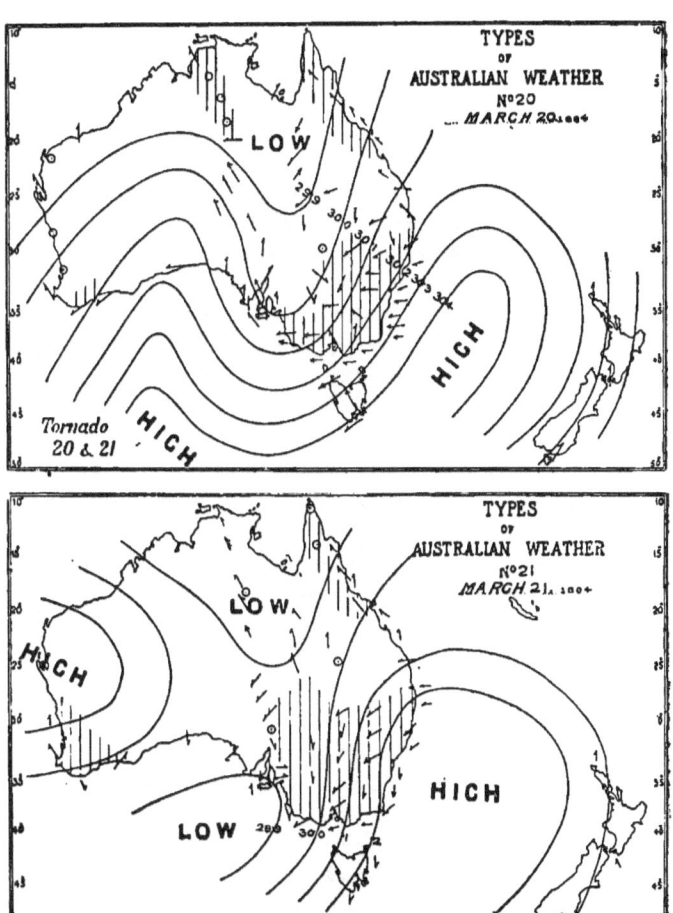

nounced then has retreated, and is now only represented by one isobar, and where the western portion of the anticyclone existed on the 20th, an antarctic depression has protruded itself, and lastly an anticyclone is entering Western Australia. Thus a most favourable and extensive col area exists between these four systems.

F

for the generation of these storms. The winds are generally blowing any way, and excepting those of the eastern high pressure have apparently no circulating power.

These were the conditions at 9 a.m., but as the day advanced the col area advanced with its arid heat, and this acting upon the precipitated moisture of the previous two days, resulted in the tornado which we are about to describe, and which occured at Bourke on the morning of March 21st, being one of the most terrific ever witnessed in that district. It struck the town at 10 o'clock, but could be seen approaching for some time from a north-westerly direction. It only lasted six minutes, but during that period thirty-nine points of rain fell, and several hundred pounds' worth of damage was done to houses. Chimneys, verandahs, trees, etc., suffered and general consternation prevailed. Many narrow escapes occurred, as the cyclone came across the common from the direction of Fort Bourke. The theatre was unroofed and a quantity of beams and iron was deposited in an adjoining yard. Pleasure boats on the river were sunk, and a steam-boat was considerably damaged by the falling of a large gum tree.

TYPE X.—CYCLONES FROM NORTH-EAST.

A comparatively small number of these storms reach the coast of Australia, and owing to the almost complete absence of observing stations, New Caledonia excepted, and the small number of vessels passing their tracks, it is usually impossible to trace their course before they reach Australia, but there seems to be no reason to doubt that they are more or less spent tropical cyclones, which reach Australia in the act of recurving. The majority reach the coast of Queensland between latitudes 20° and 26°; some farther north and south; only one has been traced from the coast inland, and then recurving there to south-east. It reached the coast in the neighbourhood of Brisbane in January 1893, passed inland over the mountains, gradually curving to south past Mudgee and Dubbo, thence curving easterly it left the mainland about latitude 35° S. Its course was marked by violent cyclonic wind and rain.

Picton reported 4·76″ rain. It has not been possible to present this as an example, because one of the days was a Sunday when very few observations are recorded.

As a rule these north-east cyclones recurve at the coast line. Charts 22 and 23 show the weather conditions on March 10th and 11th, 1891.

On March 10th there was a sudden fall of pressure on the coast of Queensland, about latitude 24° (Chart 22), and an accession

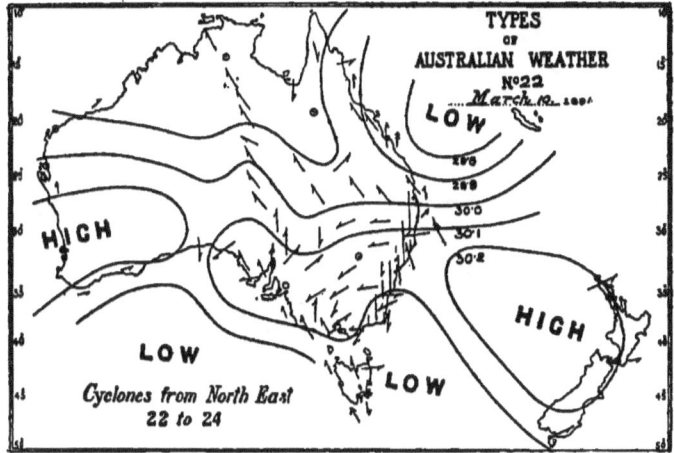

of wind and sea which clearly heralded the coming storm. This was intensified on March 11th, (Chart 23) by a further fall of three-tenths of an inch, and three more isobars of the cyclone were marked on the coast south to south-east; gales prevailed in its neighbourhood with very heavy rains.

On March 12th, (Chart 24) the anticyclone had retreated to New Zealand, and the cyclone travelled to the south along the coast intensifying as it came. The barometer curve at Brisbane was exactly of the cyclone type and dropped to 29·5. In the southern part of Queensland and northern of New South Wales heavy southerly to easterly gales were experienced. On the 13th the storm had disappeared to the eastward.

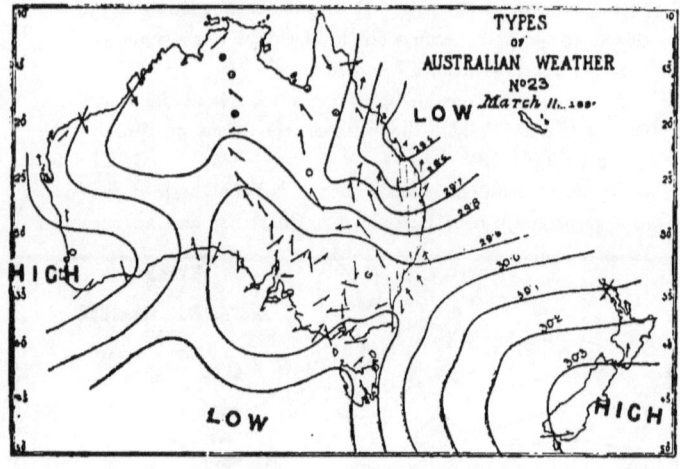

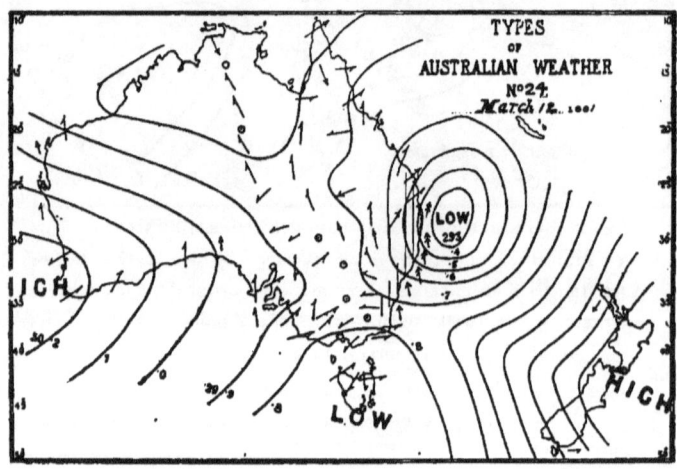

TYPE XI.—SOUTH-EAST GALES.

The south-east gale here referred to is peculiar to the east coast of Australia, and it has been responsible for the most memorable wrecks on this coast. For the most part these gales appear to be partially spent cyclones, which come in from north-east or east, and travel down the coast until they begin to recurve to the eastward.

The warning of their coming is usually very short; it consists of a sudden increase in the sea on some northern part of the coast with wind from east to south, and falling barometers, while the high pressure over Victoria and South Australia becomes intensified and progresses into the Tasman Sea. The south-east circulation about this anticyclone increases in force with the increasing barometric grade, and also by the wind circulation about the cyclone, and the effect of the two causes acting together is to produce a most serious gale. Rarely, these storms originate in a monsoonal depression somewhere over South Australia, which travelling eastward intensifies on the east coast. Heavy rain is a marked feature of these storms, but it is confined to the coast, and rarely if ever extends inland.

The storm selected to illustrate this type was a very severe one, and began on September 23rd, 1892, about 6 p.m. The barometric conditions antecedent to it are shown in Chart 25; the main

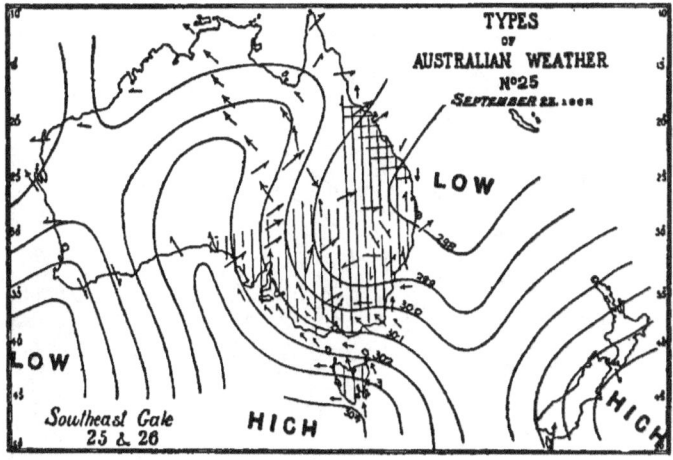

features being the bend in the high pressure isobars, and the
dormant low pressure off the coast of Queensland.

At 9 a.m. on this day there was nothing in the local weather
conditions which would lead one to anticipate the gale that
eventuated, the winds being generally light, and at Sydney only
a light breeze was blowing from the South-south-west; but at 3
p.m. an unusual fall took place in the barometers to the north-east
of New South Wales, and the winds there were freshening generally. On preparing a chart at this hour we found the depression
was intensifying and had a cyclonic tendency; at 6 p.m. in
Sydney the wind, which had been blowing from the south-east
and gradually increasing in velocity, reached the force of a gale,
a thick driving rain began to fall, which continued with little intermission until daybreak next day, when over 2" were registered,
and an extensive area around the metropolis benefitted to the
extent of an inch and upwards; the barometer at this hour also
began to fall rapidly and steadily, until at 5 a.m. on the 24th it
read 29·203, and the wind had reached in one squall, lasting only
a few seconds, the extraordinary rate of one hundred and twenty
miles per hour, the mean rate of the gale being thirty-two miles
per hour.

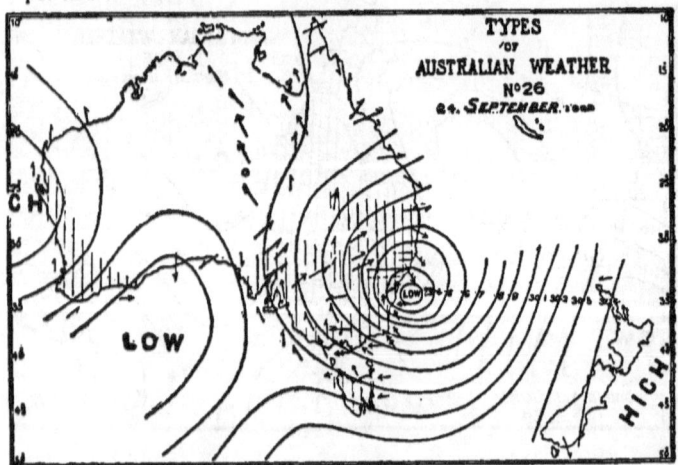

At 9 a.m. on the 24th, (Chart 26) the cyclone was still in an active state, but it had passed to the south of Sydney and was receding from the coast; the barometers were rising rapidly on shore, and before noon a light northerly wind was blowing. On the 25th September, the day following the gale, the weather everywhere was generally fine, while all that remained of the energetic high and low pressure systems were parallel isobars lying over the southern areas of Australia with very shallow gradients. This gale was very destructive and did much damage to property in Sydney; in some instances houses were unroofed, and the wind and sea on the coast were very heavy.

TYPE XII.—DEVELOPMENT OF A CYCLONE FROM A Λ DEPRESSION.

The distinction between a Λ depression and a monsoonal low pressure is not by any means well defined, and it is possible that this should be taken as a variation of Type 3; there are however, marked differences, not only in the shape of the isobars, but also in the wind; and the most decided distinction is perhaps the easterly wind circulating round the southern part of the monsoonal low pressure and the northerly and southerly winds about the Λ; but in some cases, as in the one selected, the wind circulation is mixed, northerly, southerly and easterly winds being present, and these from their want of energy tend to throw the forecaster off his guard. The season is, however, some guide, as these storms are most frequent from September to April.

Their sphere of influence is very extensive, as may be noted in Chart 28, which shows rain over half Australia as the apparent result of this storm. Most of these storms take a direct easterly course over New South Wales and Victoria or through Bass' Straits. At times they move to north-east, the Polar winds being more energetic, and this feature intensifies all the storm and rain conditions. The winds in all these storms are violent, and in some very destructive.

The one selected for illustration appeared first on 27th May, 1893, (Chart 27). At 9 a.m. on that day, a dormant and irregular

A depression existed over the Australian Bight, the winds were moderate northerly and southerly, as usual in such conditions, there were a few light easterly winds. Isobars were close over eastern Victoria, but there was nothing which seemed to indicate the violent cyclone that developed during that day over southern parts of South Australia and south-west of New South Wales. (See Charts 27 and 28.) This storm formed in the rear of a very substantial anticyclone, then over the Tasman Sea, and it should be noted that it did not act as a secondary and travel round the southern and eastern parts of the high pressure, but it moved towards the northern side of it and against its circulation, thus proving its own Polar impulse and energy, and giving rise to very strong gales and steep barometric grades with great fall of temperatures; these conditions produced extremely heavy and wide spread rains, not only within the storm isobars, but over the whole of the eastern half of Australia. We have no means of tracing the rain to the west of the overland telegraph line, because there are no observing stations there.

May 28th was unfortunately a Sunday, and we have no observations for that day, but on the 29th the cyclone is seen in full

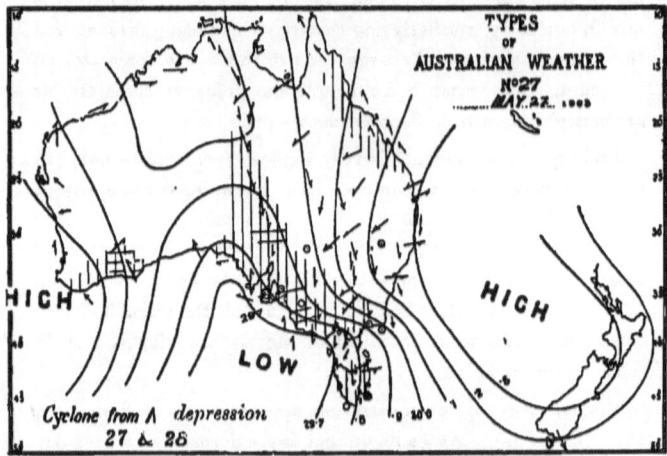

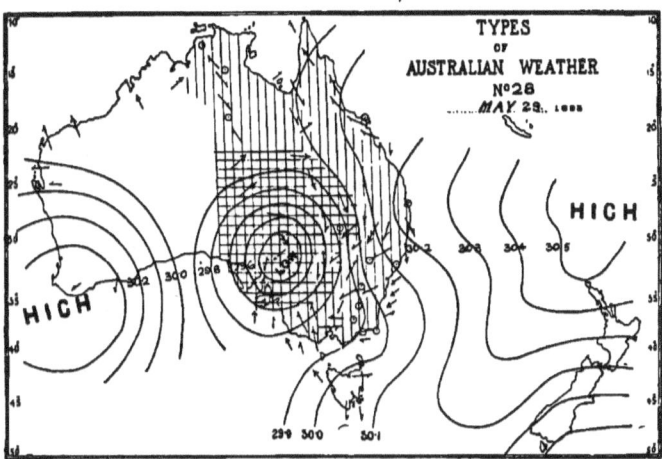

strength covering the southern colonies of South Australia and Victoria, and the greater part of New South Wales; its isobars are unusually symmetrical and its rain influence the most extensive we have on record. During the 29th and 30th rain continued to fall over considerable areas, although on the 30th the depression filled up and the storm was displaced by the high pressure coming on from the west.

TYPE XIII.—WESTERLY WINDS.

The Winter anticyclone (See Chart 36) is much more extensive than the Summer one, and its grades are steeper and circulation stronger, while its latitude is further north, often up to 30° S.; hence the circulation on its southern side, unlike the Summer one, affects the mainland of Australia and gives us our westerly winds. And just as the trade wind intensifies its northern circulation by adding force thereto, so the brave west winds of the southern ocean follow the general move of the weather systems northwards, and thus add force to the westerly circulation of the anticyclone, and its greater dimensions increase the size of the Λ depression, so much that it no longer has the sharply defined change from notherly to southerly winds, but from north-west to south-west winds; all these conditions combine to strengthen the westerly circulation, until at times they seem to be like the brave west winds of the roaring forties.

On the east coast they extend further north than they do inland, and at times include Brisbane within their influence. When very strong the anticyclone is elongated and the southern isobars are flattened. (See Chart 30). They are very cold and dry, having all the raw feeling of the easterly winds of England, and have been known to last for several weeks without intermission; they are the most persistent winds during our winter. In heavy westerly winds the cold is very severe in Bass' Straits, between the south coast of Victoria and Tasmania, rain, hail, and sleet being very common. It is almost needless to say that on land they bring very little rain, and they rapidly dry up the soil. The severe drought of 1895 was largely due to the persistent westerly winds which rapidly dried up the occasional rains; this feature becomes intensified if instead of coming from due west they blow from north-west.

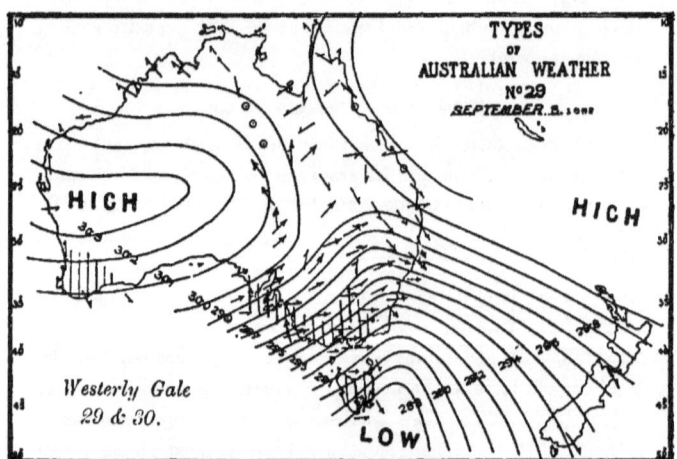

Turning now to the charts of the westerly gale selected for illustration, it will be seen in Chart 29, that this storm began on September 3rd, 1895; on that day an elongated anticyclone lay over Western Australia, a flattened and extensive ʌ over New South Wales and Tasman Sea, and the winds generally displayed great energy, as might be expected from the close isobars, and unusually low barometers over Tasmania; light rain was falling on the coasts of South Australia and Victoria.

On September 4th, Chart 30, the anticyclone is more elongated and the ʌ flattened until its isobars are nearly horizontal, and heavy westerly gales swept all the south-eastern part of Australia and all Tasmania. On the 3rd the wind at Sydney at noon for a short time reached a velocity of seventy-eight miles per hour. On the 4th the wind was less gusty, but its average velocity was quite as strong as it was on the 3rd.

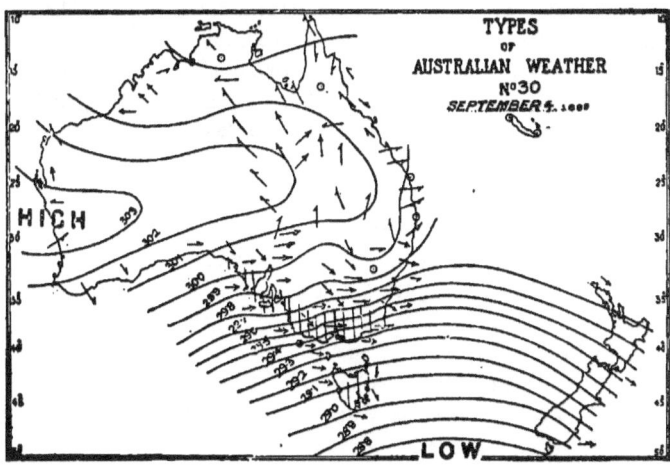

TYPE XIV.—SOUTHERLY BURSTERS.

The southerly burster is a well known feature or type of Australian weather, so well marked in character indeed that it requires no special training in meteorology to recognise it; its characteristics are so obtrusive that they cannot be overlooked, and they are welcomed as the most pleasant relief after the high temperatures and oppressive northerly winds which precede them. They come in the late spring, all summer, and part of autumn, but as a rule, in order to get a strong southerly an antecedent excessively hot day or days must be experienced. The duration of a southerly burster may be anything from two hours to ten days, but it is not to be understood that the term burster is applied to the whole period or duration of the southerly wind. What is called a burster is the squall or sudden and violent change of wind direction, and the violent rush or " burst " which marks the advent of this wind. We need not go into all the characteristics; these will be found in the Abercromby Prize Essay on this subject.[1] It is there explained that the south wind comes in front of an approaching anticyclone, and that it is felt from West to Eastern Australia, but it is only on the eastern coast, where, aided by the smaller friction of the ocean and the shelter which the mountains afford from other winds, that the southerly becomes more vigorous and rushes northwards in a squall, which happens so suddenly and with such force, that at times ships drag their anchors in Sydney harbour.

It is not definitely made out yet that these storms are ever "line storms" in the sense that the change of wind comes as the dip in the isobars passes over each place in succession, but there are many facts which suggest that such is the fact in some instances. Our present purpose is to describe a "burster" as a type of Australian weather. The essential feature of it is a sharp ʌ; in Chart 31 such a depression is shewn existing over Victoria and Tasmania, with its axis lying from north-north-west to south-south-east. An anticyclone of good energy for this time of the year exists to the west, and hot northerly winds occupy northern Australia; these are the elements for the good burster that followed. As a general rule,[2] the position and character of two

---

[1] Journal Royal Society, N. S. Wales, Vol. XXVIII.
[2] See Moving Anticyclones, p. 4.

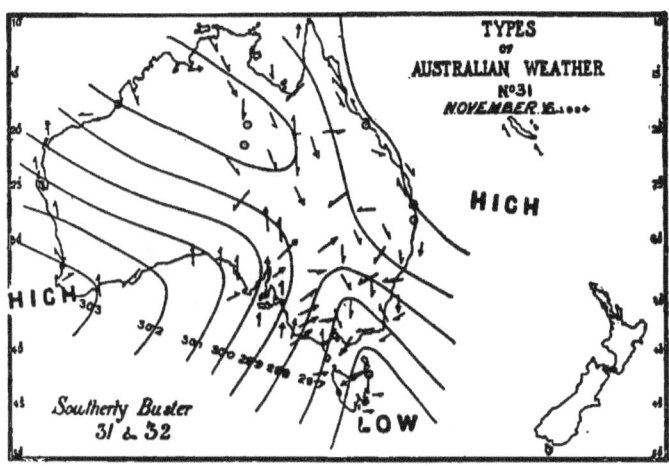

such systems as shown would bring the burster to the coast of New South Wales within twenty-four hours; in this case it took thirteen hours.

November 16th, light to fresh north-east winds were blowing on the coast at 9 a.m., while in the front of the high pressure strong south-west winds were blowing. In Chart 32, the southerly

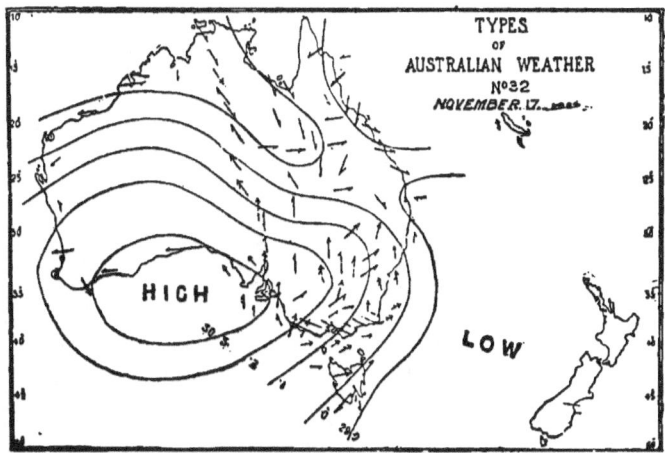

is shown in full force in front of the anticyclone, which by the way has lost none of its energy since the previous day; the depression is well off the coast and on its way to New Zealand. This burster reached a velocity of forty-nine miles per hour, and lasted thirty-five hours.

TYPE XV.—THE BLACK NORTH-EASTER.

This is a somewhat uncommon but nevertheless well known type of weather on the coast about Sydney. Its characteristics are a very strong and persistent north-east gale, continuing day and night for two or more days; it has been known to last five days and nights, and it ends with the advent of a southerly burster. Its cause is found in an extensive col, the rear of one anticyclone being at rest over this coast, while another lies over the Australian Bight. If the grade is rather steep and the system at rest for several days, then the north-east wind persists with force proportionated to the grade, until the whole system moves forward; the southerly winds in the front of the approaching high pressure then displace the north-easter and the storm is over.

There have been no good examples of this type since weather charts have been printed here. Chart 33 shows the necessary forms of isobars, but the grade is not steep enough for a gale.

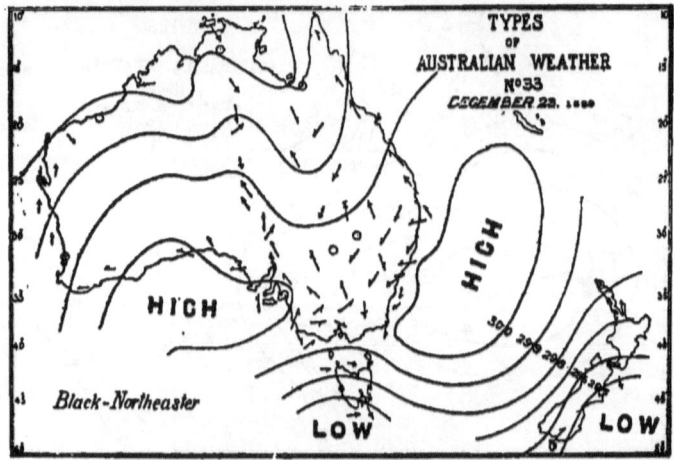

TYPE XVI.—WIND BLOWING CONTRARY TO ISOBARS.

In this type the wind blows with considerable force in a direction directly opposed to that which the isobars would lead us to expect. For instance in No. 34, it will be seen that an extensive high pressure rests over the east coast, and the isobars are comparatively close together. The normal circulation with these isobars would be fresh north to north-west winds, when, as a matter of fact, strong southerly winds were blowing as far as Sydney with a velocity of thirty miles per hour. Such conditions are rather troublesome in forecasting; fortunately they do not come often, and the fact is not confined to southerly winds. The general direction of the coast line is northerly bearing east a little, a range of mountains from two to four thousand feet high runs nearly paralled to it, and this local formation has a very important effect on the circulation of the wind; as in Chart 34 it seems to have more effect than the isobars, and probably the grade was rapidly intensifying and had not been long enough in existence to fully control the winds. It seems probable, so far as this type has been studied, that we should find that when the wind blows contrary to isobars, it does so because of some impulse given to it before the new grade had time to control the circulation.

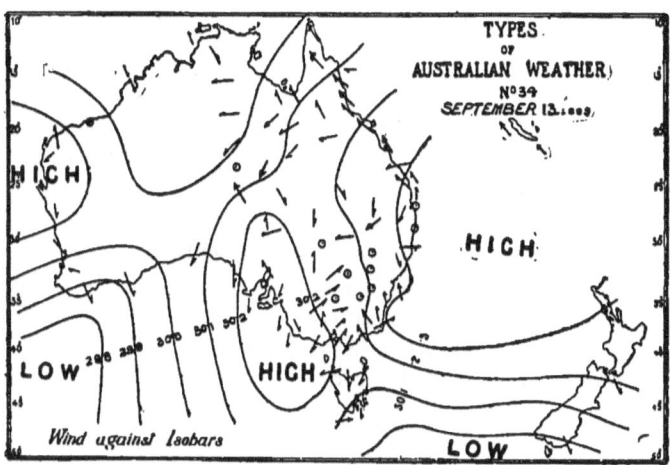

TYPE XVII.—SUMMER ANTICYCLONES.

One of the most marked features of weather in Australia is the regular easterly motion of it in all seasons of the year, added to this the anticyclones—the controlling element in our weather—change with the sun, so that the latitude they follow in their easterly progress is further south in summer than in winter.[1] The summer latitude is about 40° south. They are less extensive in summer than in winter, and do not so completely control our weather as they do in winter, for their southern position leaves room for the southerly extension of monsoonal low pressures, which make a great deal of our weather; but the anticyclone makes the change from hot northerly to cool southerly winds, the bursters of the east coast.

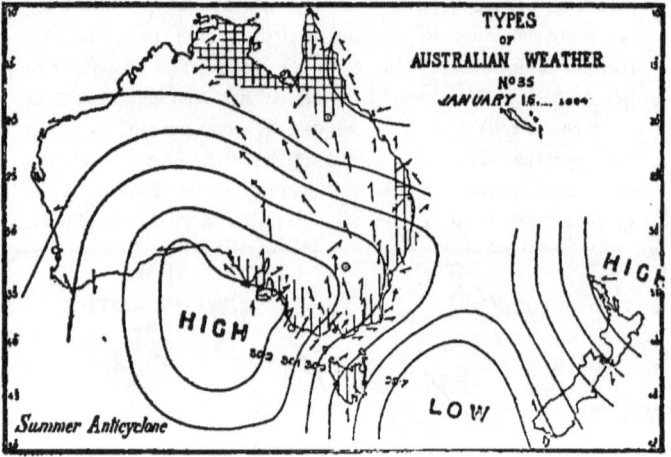

TYPE XVIII.—WINTER ANTICYCLONES.

During the winter months the antiycclones are much larger than they are in summer, and their latitude about 30° S.; very commonly their area is equal to Australia, and their control of the weather more complete than it is in summer. Fine weather

---

[1] See Moving Anticyclones, page 2.

marks their centres, and the rains come chiefly from the low pressures between each pair of anticyclones, and the strong westerly gales are but part of the circulation about these high pressures. (See Charts 29 and 30.)

The one selected for illustration occupied the whole of Australia on June 4th, 1895, Chart 36. Its form was remarkably symmetrical, and in the central area the barometers read 30·6, which is somewhat unusual, hence the circulation is active, and in northern Australia where the trade wind adds force, it is very strong. Under the central influence of these great anticyclones the whole of Australia enjoys fine weather.

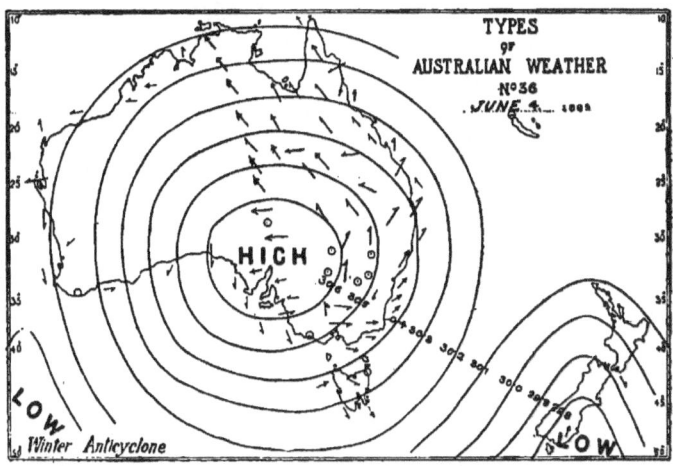

TYPE XIX.—SQUARE HEADED A DEPRESSION.

This type is a variation of the usual ʌ depression, but is sufficiently characteristic to be placed by itself as a type or rather a sub-type. Its isobaric peculiarity, as may be seen by reference to Charts 37 and 38, is that there is a flat top or square head to the isobars of the ʌ, the usual form of which is a sharp and regular curve, and the marked feature of this is that under it the weather is remarkably squally and charged with thunder and hail storms, and deluges of rain. This type is not peculiar to any season of the year, and the country they affect is usually south of Lat. 30°. The particular square headed Λ depression under discussion occurred in July, 1891. Its effects were widespread and its life persistent, and its peculiarity in a more or less distinctive form was maintained during its passage over six degrees of longitude, and its effects most severe in Victoria.

In Chart 37, July 8th, 1891, the square headed Λ is shown when over South Australia. The gradients about it excepting those on its eastern side were moderate; winds were from fresh to strong. Following it is a high pressure of decided energy, and preceding it to the east an anticyclone of little or no character; on the previous day rain fell over greater part of the southern seaboard and in Tasmania.

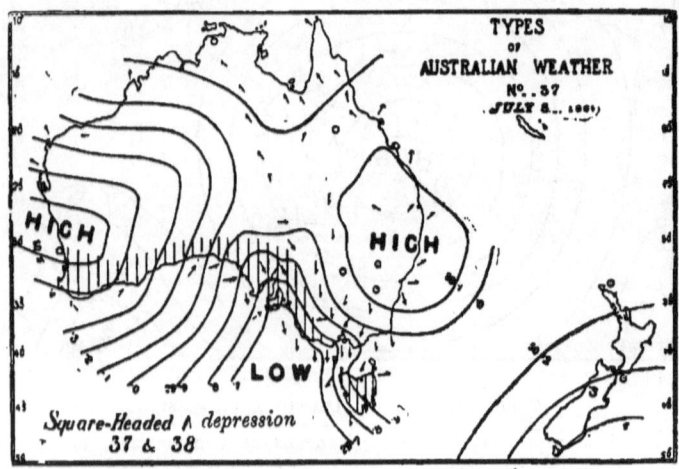

Chart 38, July 9th, the depression has become more extensive, but has lost energy. The relative powers of the anticyclone remains the same, winds were generally lighter, but the rain covered a much wider area than on the previous day. As this storm moved to the east its characteristic weather was maintained.

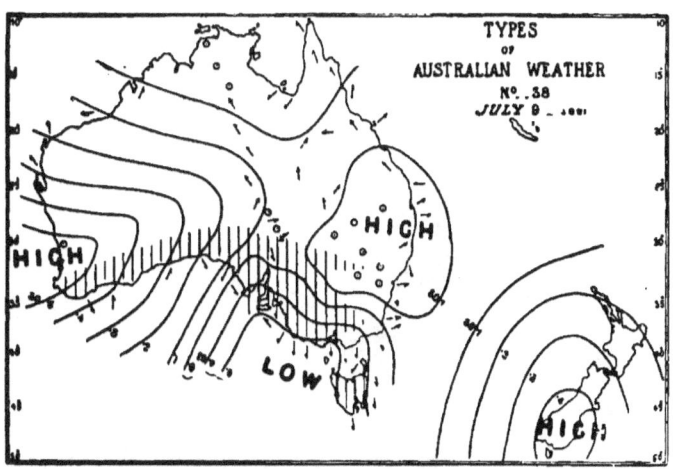

TYPE XX.—THE ADVENT OF AN ANTARCTIC STORM.

This is a type of weather that does not often visit Australia but its severity makes it noteworthy for the winds and weather which come with it are very destructive, and the cold severe, it might almost be called a southern blizzard; yet the warning is short and often difficult to read, for it comes from the Antarctic where we have no out stations.

The storm under consideration began to affect the south coast of Australia, making the winds fresh to strong on June 21st, 1892, (Chart 39), yet the season (winter) and the general conditions pointed to westerly winds, and said nothing definite of the storm which was telegraphed on the morning of June 22nd, (Chart 40). Barometers were then seen to be four-tenths lower in Tasmania than they were on the 21st, and the wind had increased to a furious gale along the south coast of South Australia and Victoria, and the isobars indicate a very steep grade commensurate with the wind. On shore also the wind rose in places to hurricane force; at Ballarat, in western Victoria, buildings were unroofed and trees blown down. So severe was it there that the storm is recorded as the Ballarat storm.

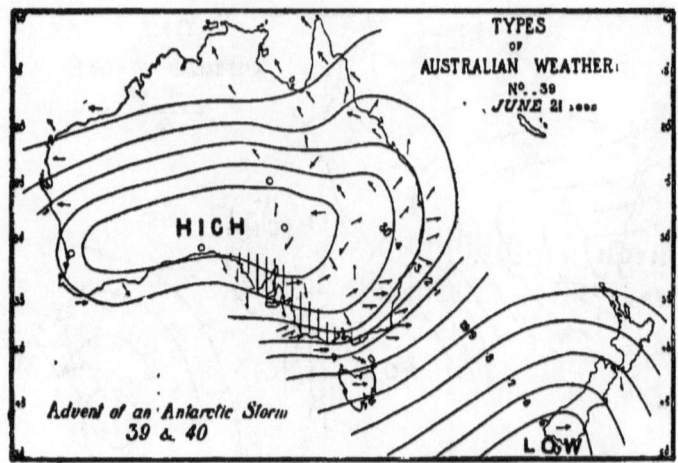

In South Australia and New South Wales the wind did a great deal of damage without reaching the intensity it had in Victoria. The isobars show a retreat of the anticyclone and a tilting of the major axis, and a very remarkable increase in the number of isobars on the south coasts of South Australia, Victoria, and

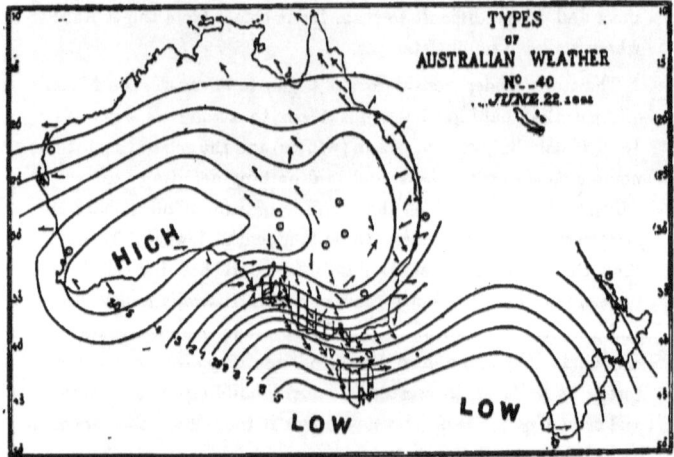

Tasmania. It looks as if the low pressure had retreated and its western parts forced their way north. What probably did take place was that a storm centre south of the Australian Bight and indicated by the northerly winds in Chart 39, had in the interval surged northwards on to the west coast of Victoria, bringing with it all its antarctic energy and severe cold. This view is supported by the fact that there was in the twenty-four hours but little change in the New Zealand isobars, and further by the upward tilting of the eastern part of the anticyclone caused by the northing of the antarctic storm, and lastly by the blizzard-like cold which was so marked a feature of this storm.

LIST OF TYPES OF AUSTRALIAN WEATHER.

I.—Moving Anticyclones, Charts 1, 2, 3.
II.—Monsoonal Rain Storm, Charts 4, 5.
III.—Development of a Cyclonic Storm in Low Latitudes from a Monsoonal Depression, Charts 6, 7.
IV.—Development of a Cyclonic Storm in High Latitudes from a Monsoonal Depression, Charts 8, 9, 10.
V.—Conditions favourable for Thunderstorms, Charts 11, 12.
VI.—Cyclonic Thunderstorms, Charts 13, 14.
VII.—Vertical and nearly straight Isobars, Charts 15, 16.
VIII.—Cyclones from North-West, Charts 17, 18, 19.
IX.—Cyclones from North-East, Charts 20, 21, 22.
X.—Tornadoes, Charts 23, 24.
XI.—South-East Gales, Charts 25, 26.
XII.—Development of Cyclones from a ʌ Depression, Charts 27, 28.
XIII.—Westerly Winds, Charts 29, 30.
XIV.—Southerly Bursters, Charts 31, 32.
XV.—Black North-Easter, Chart 33.
XVI.—Winds Blowing Against Isobars, Chart 34.
XVII.—Summer Anticyclone, Chart 35.
XVIII.—Winter Anticyclone, Chart 36.
XIX.—Square Headed ʌ Depression, Charts 37, 38.
XX.—Advent of an Antarctic Storm, Charts 39, 40.

Sydney:
F. W. WHITE, PRINTER, 39 MARKET STREET.
1896.

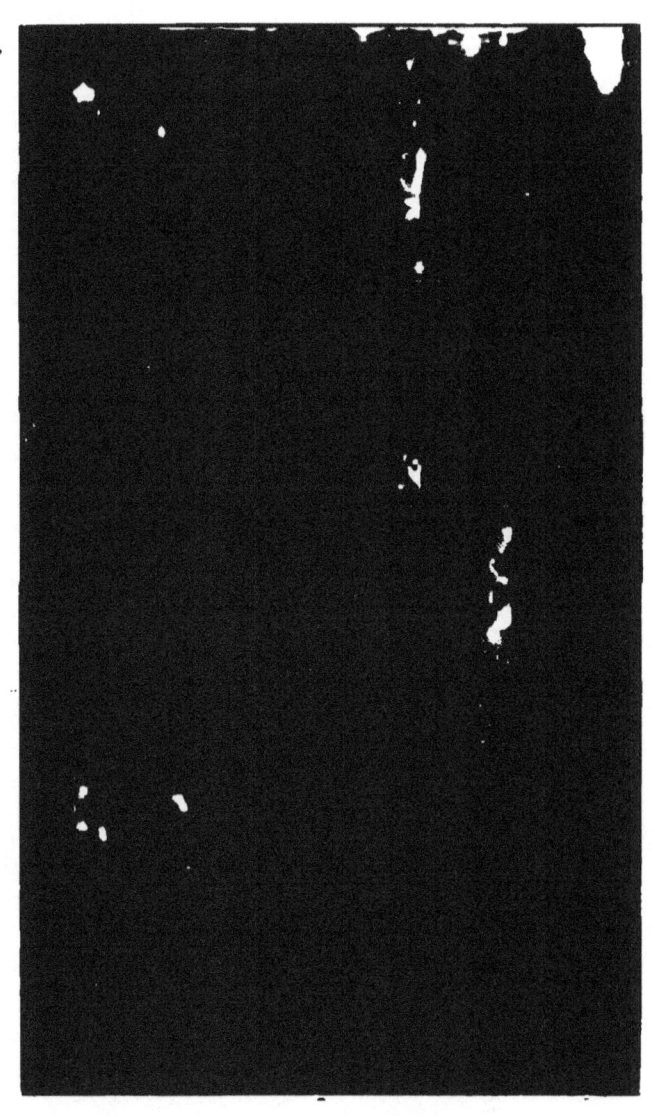

Abercromby "Southerly Bursters." Plate IV.

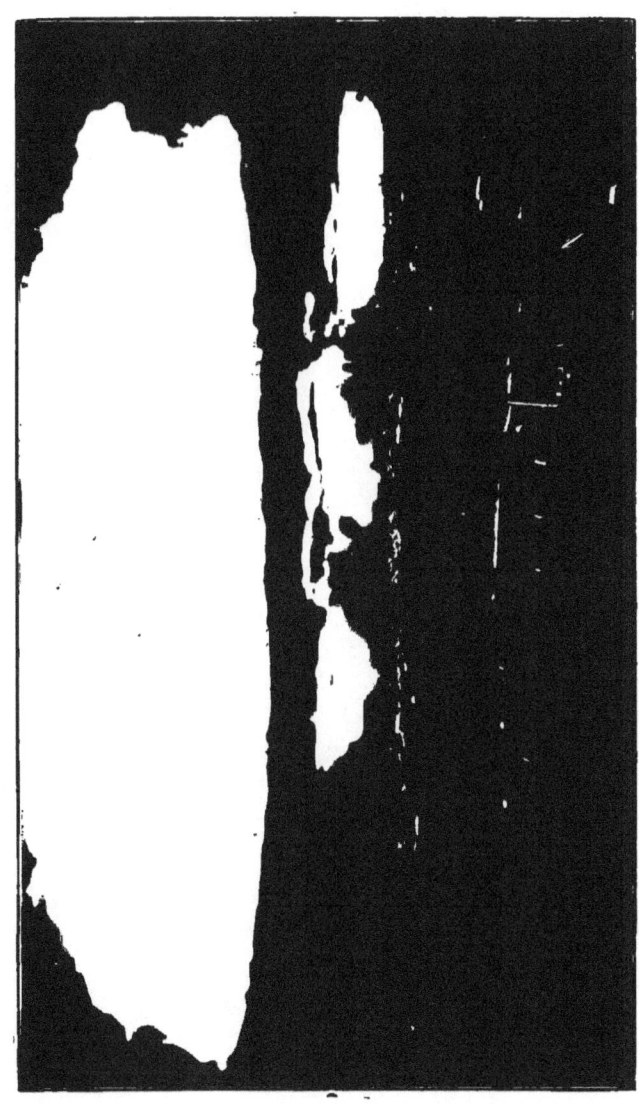

www.ingramcontent.com/pod-product-compliance
Lightning Source LLC
Chambersburg PA
CBHW020140170426
43199CB00010B/821